110 Geometry Problems

for the International Mathematical Olympiad

110 Geometry Problems

for the International Mathematical Olympiad

Titu Andreescu

Cosmin Pohoata

Library of Congress Control Number: 2014915529

ISBN: 978-0-9885622-2-6

9 8 7 6 5 4 3 2 1

www.awesomemath.org

Cover design by Iury Ulzutuev

Preface

This book is an unofficial sequel to the first two geometry books published by XYZ Press, namely 106 Geometry Problems from the AwesomeMath Summer Program and 107 Geometry Problems from the AwesomeMath Year Round Program. Assuming the background presented in these two books, 110 comes as a collection of problems designed for passionate geometers and students preparing for the difficult geometry questions of the IMO.

We know by now that the key to problem solving in geometry is mastering the basics and being able to apply them efficiently. Nonetheless, not all geometry problems can be solved by simply building up from nineteenth century geometry. Some results require deep insight into the configurations, as well as insight that goes well beyond the naive pursuit of common sense techniques. The book we have written is a collection of our favorite such problems.

The solutions and the commentaries that usually follow them represent however the heart of our work. Since a non-trivial portion of the problems were in fact proposed by us in various contests around the world, we often chose to reproduce the way we initially thought about them as the way to present the proofs. This is not only meant to give the reader a glimpse into how authors come up with their problems, but also to induce a structural way of thinking when it comes to intricate configuration debugging. It is also the way we came up with most of the solutions for the non-personal problems that you will encounter in this book.

We wish you a pleasant reading.

Titu Andreescu and Cosmin Pohoata

Abbreviations and Notation

Notation of geometrical elements

$\angle BAC$	convex angle by vertex A
$\angle(p,q)$	directed angle between lines p and q
$\angle BAC \equiv \angle B'AC'$	angles BAC and $B'AC'$ coincide
AB	line through points A and B, distance between point A and B
$\overline{AB}$	directed segment from point A to point B
$X \in AB$	X lies on the line AB
$X = AC \cap BD$	X is the intersection of the lines AC and BD
$\triangle ABC$	triangle ABC
$[ABC]$	area of $\triangle ABC$
$[A_1 \dots A_n]$	area of polygon $A_1 \dots A_n$
(ABC)	circumcircle of triangle ABC
$(A_1 \dots A_n)$	area of cyclic polygon $A_1 \dots A_n$
$AB \| CD$	lines AB and CD are parallel
$AB \perp CD$	lines AB and CD are perpendicular
$p(X,\omega)$	power of point X with respect to circle ω
$\triangle ABC \cong \triangle DEF$	triangles ABC and DEF are congruent (in this order of vertices)
$\triangle ABC \sim \triangle DEF$	triangles ABC and DEF are similar (in this order of vertices)
$\mathcal{H}(H,k)$	homothety with center H and factor k
$\Psi(P,r^2)$	inversion with center P and power r^2
$\mathcal{S}(S,k,\varphi)$	spiral similarity with center S, dilation factor k, and angle of rotation φ

Notation of triangle elements

a, b, c	sides or side lengths of $\triangle ABC$
$\angle A, \angle B, \angle C$	angles by vertices A, B, and C of $\triangle ABC$
s	semiperimeter
x, y, z	expressions $\frac{1}{2}(b+c-a)$, $\frac{1}{2}(c+a-b)$, $\frac{1}{2}(a+b-c)$
r	inradius
R	circumradius
K	area
h_a, h_b, h_c	altitudes in $\triangle ABC$
m_a, m_b, m_c	medians in $\triangle ABC$
l_a, l_b, l_c	angle bisectors (segments) in $\triangle ABC$
r_a, r_b, r_c	exradii in $\triangle ABC$

Abbreviations

USAMO	United States of America Mathematical Olympiad
USAJMO	USA Junior Mathematical Olympiad
IMO	International Mathematical Olympiad
Romania IMO TST	Romania IMO Team Selection Test
USA IMO TST	United States of America IMO Team Selection Test
RMM	Romanian Masters of Mathematics

Contents

Problems

Problem 1. Given two nonintersecting chords AB and CD in a circle and a variable point P on the arc AB remote from points C and D, let E and F be the intersections of chords PC, AB, and of PD, AB, respectively. Prove that the value of

$$\frac{AE \cdot BF}{EF}$$

does not depend on the position of P.

Problem 2. Consider a triangle ABC with $a \leq b \leq c$. Denote by X, Y, Z the midpoints of the sides BC, CA, and AB, respectively. Let D, E, F be points on the sides BC, CA, AB satisfying the following two conditions:

(i) D is between X and C, E is between Y and C, and F is between Z and B.

(ii) $\angle CDE \leq \angle BDF$, $\angle CED \leq \angle AEF$, and $\angle BFD \leq \angle AFE$.

Prove that the perimeter of triangle DEF does not exceed the semiperimeter of triangle ABC.

Problem 3. Let A' be the foot of the internal angle bisector of the angle BAC of a given triangle ABC. Let P be an arbitrary point on the cevian AA', different from A', and denote by B', C' the intersections of the lines BP, CP, with the sidelines CA, and AB, respectively. If $BB' = CC'$, prove that $AB = AC$.

Problem 4. Let ABC be a triangle and let A', B', C' be the feet of the altitudes from A, B, C, respectively. Let A_1 be the foot of the perpendicular from A' to AB and let A_2 be the foot of the perpendicular from A' to AC. Furthermore, let Ω_A be the circle centered at vertex A having radius AA'. Analogously, define points B_1, B_2, C_1, C_2, and circles Ω_B and Ω_C. Prove that:

(a) Points A_1, A_2, B_1, B_2, C_1, C_2 are all concyclic.

(b) The center of the circle from (a) has equal powers with respect to Ω_A, Ω_B and Ω_C.

Problem 5. Let $\mathcal{C}$ be a circle and let P be a point in the exterior of $\mathcal{C}$. The tangents from P intersect the circle at A and B, respectively. Let M be the midpoint of segment AP and N the second intersection of the line BM with circle $\mathcal{C}$. Prove that $PN = 2MN$.

Problem 6. Let ℓ be a line in the exterior of a given circle $\rho(O)$. Let A be the foot of the perpendicular from O to ℓ, and let M be an arbitrary point on ρ. Furthermore, let X and Y be the second intersections of the circle with diameter AM with ρ and ℓ, respectively. Prove that the line XY passes through a point which is independent of the position of M.

Problem 7. Let X, Y, Z be the midpoints of the arcs BC, CA, AB, respectively, of triangle ABC containing the vertices of the triangle. Prove that the Simson lines of X, Y, Z with respect to ABC are concurrent.

Problem 8. Let ABC be a triangle with circumcenter O and let D, E, F be any three points lying on the sides BC, CA, AB, respectively. Let D', E', F' be the reflections of D, E, F with respect to the midpoints of BC, CA, AB, respectively. Prove that

(a) The Miquel points M of D, E, F and M' of D', E', F' are equidistant from the circumcenter of ABC.

(b) The centroids of triangles DEF and $D'E'F'$ are symmetric with respect to the centroid of ABC.

(c) Triangles DEF and $D'E'F'$ have the same area.

Problem 9. Let ABC be a triangle with $\angle BAC < \angle ACB$. Let D, E be points on the sides AC and AB, respectively, such that the angles ACB and BED are congruent. If F lies in the interior of the quadrilateral $BCDE$ such that the circumcircle of triangle BCF is tangent to the circumcircle of DEF and the circumcircle of BEF is tangent to the circumcircle of CDF, prove that the points A, C, E, F are concyclic.

Problem 10. Let γ be a circle and l a line in its plane. Let K be a point on l, located outside of γ. Let KA and KB be the tangents from K to γ, where A and B are distinct points on γ. Let P and Q be two points on γ. Lines PA and PB intersect line l in two points R and respectively S. Lines QR and QS intersect the second time circle γ in points C and D. Prove that the tangents from C and D to γ are concurrent on line l.

Problem 11. Given a scalene acute triangle ABC with $AC > BC$ let F be the foot of the altitude from C. Let P be a point on AB, different from A so that $AF = PF$. Let H, O, M be the orthocenter, circumcenter and midpoint of side AC. Let X be the intersection point of BC and HP. Let Y be the intersection point of OM and FX and let OF intersect AC at Z. Prove that F, M, Y, Z are concyclic.

Problem 12. Let ABC be an arbitrary triangle and let I be the incenter of ABC. Let D, E, F be the points on lines BC, CA, AB respectively such that $\angle BID = \angle CIE = \angle AIF = 90°$, and define the following measurements: r_a, r_b, r_c are the exradii of triangle ABC, $[DEF]$ is the area of DEF, and $[ABC]$ is the area of ABC. Prove that

$$\frac{[DEF]}{[ABC]} = \frac{4r(r_a + r_b + r_c)}{(a+b+c)^2}.$$

Problem 13. Let a, b, and c be the lengths of the sides opposite vertices A, B, and C, respectively, in a nonobtuse triangle. Let h_a, h_b and h_c be the corresponding lengths of the altitudes. Show that

$$\left(\frac{h_a}{a}\right)^2+\left(\frac{h_b}{b}\right)^2+\left(\frac{h_c}{c}\right)^2\geq\frac{9}{4},$$

and determine the cases of equality.

Problem 14. Let ABC be an acute-angled triangle with orthocenter H, and let W be a point on the side BC. Denote by M and N the feet of the altitudes from B and C, respectively. Denote by ω_1 the circumcircle of BWN, and let X be the point on ω_1 which is diametrically opposite to W. Analogously, denote by ω_2 the circumcircle of triangle CWM, and let Y be the point on ω_2 which is diametrically opposite to W. Prove that X, Y and H are collinear.

Problem 15. Let ABC be a triangle and let P be a point in its interior. Let X, Y, Z be the intersections of AP, BP, CP with sides BC, CA, AB, respectively. Prove that

$$\frac{XB}{XY}\cdot\frac{YC}{YZ}\cdot\frac{ZA}{ZX}\leq\frac{R}{2r}.$$

Problem 16. Let $\mathcal{C}_1$ be a circle that is tangent to sides AB and AC of triangle ABC and let $\mathcal{C}_2$ be a circle through B and C that is tangent to $\mathcal{C}_1$ at point D. Prove that the incenter of triangle ABC lies on the internal angle bisector of $\angle BDC$.

Problem 17. Let ABC be a triangle and let P, Q be two points lying on its circumcircle. Prove that their Simson lines meet on the A-altitude of triangle ABC if and only if $PQ\|BC$.

Problem 18. Let ABC be a triangle and let P be a point in its interior with pedal triangle DEF. Suppose that that the lines DE and DF are perpendicular. Prove that the isogonal conjugate of P with respect to ABC is the orthocenter of triangle AEF.

Problem 19. Let τ be a line that is tangent to the circumcircle $\Gamma(O,R)$ of triangle ABC. Let I, I_a, I_b, I_c be the incenter and the three excenters of ABC and let $\delta(P)$ stands for the distance from point P to τ. Prove that there exists a choice of signs so that the following identity is true:

$$\pm\delta(I)\pm\delta(I_a)\pm\delta(I_b)\pm\delta(I_c)=4R.$$

Problem 20. Prove that the nine-point circle of a triangle is tangent to the incircle and the excircles of the triangle.

Problem 21. In acute triangle ABC, $\angle A < \angle B$ and $\angle A < \angle C$. Let P be a variable point on side BC. Points D and E lie on sides AB and AC, respectively, such that $BP = PD$ and $CP = PE$. Prove that as P moves along side BC, the circumcircle of triangle ADE passes through a fixed point other than A.

Problem 22. Let ABC be a triangle with circumcenter O and let X, Y, Z be the circumcenters of triangles BOC, COA, AOB. Prove that the lines AX, BY, CZ are concurrent.

Problem 23. Let ABC be a triangle and let P be a point in its interior. Let the line passing through P, which is perpendicular to PA, intersect BC at A_1. Analogously, define B_1 and C_1. Prove that A_1, B_1, C_1 are collinear.

Problem 24. Given a quadrilateral $ABCD$ with $\angle B = \angle D = 90°$. Point M is chosen on segment AB so that $AD = AM$. Rays DM and CB intersect at point N. Points H and K are the feet of the perpendiculars from D and C to lines AC and AN, respectively. Prove that $\angle MHN = \angle MCK$.

Problem 25. Let P be a point in the plane of a triangle ABC and let Q be its isogonal conjugate with respect to ABC. Prove that

$$\frac{AP \cdot AQ}{AB \cdot AC} + \frac{BP \cdot BQ}{BA \cdot BC} + \frac{CP \cdot CQ}{CA \cdot CB} = 1.$$

Problem 26. In a triangle ABC, let A_1, B_1, C_1 be the points where the excircles touch the sides BC, CA and AB respectively. Prove that AA_1, BB_1 and CC_1 are the side lenghts of a triangle.

Problem 27. Prove that the three points of intersections of the adjacent internal angle trisectors of a triangle form an equilateral triangle.

Problem 28. Let Ω be the circumcircle of triangle ABC and let D be the tangency point of its incircle $\rho(I)$ with the side BC. Let ω be the circle internally tangent to Ω at T, and to BC at D. Prove that $\angle ATI = 90°$.

Problem 29. Let ABC be a triangle and let ℓ be a tangent to its incircle. Let x, y, z be the signed distances from A, B, C to ℓ, respectively. Prove that

$$ax + by + cz = 2[ABC],$$

where $[ABC]$ denotes the area of triangle ABC.

Problem 30. Let a, b, c, and x, y, z be the side lengths of two given triangles ABC, XYZ with areas $[ABC]$, and $[XYZ]$, respectively. Then,

$$a^2\left(y^2+z^2-x^2\right)+b^2\left(z^2+x^2-y^2\right)+c^2\left(x^2+y^2-z^2\right) \geq 16[ABC][XYZ],$$

with equality if and only if the triangles ABC and XYZ are similar.

Problem 31. Diagonals of a cyclic quadrilateral $ABCD$ intersect at point K. The midpoints of segments AC and BD are M and N, respectively. The circumcircles of triangles ADM and BCM intersect at points M and L. Prove that K, L, M and N all lie on a circle.

Problem 32. Let ABC be a triangle with incenter I, incircle γ and circumcircle Γ. Let M, N, P be the midpoints of sides BC, CA, AB and let E, F be the tangency points of γ with CA and AB, respectively. Let U, V be the intersections of EF with MN and MP, respectively, and let X be the midpoint of arc BAC of Γ. Prove that XI bisects UV.

Problem 33. Let ABC be a triangle and let M, N, P be the midpoints of sides BC, CA, AB, respectively. Denote by X, Y, Z the midpoints of the altitudes emerging from the vertices A, B, C, respectively. Prove that the radical center of the circles AMX, BNY, CPZ is the nine-point center of triangle ABC.

Problem 34. Let $A_1A_2A_3A_4$ be a quadrilateral with no pair of parallel sides. For each $i = 1, 2, 3, 4$, define ω_1 to be the circle touching the quadrilateral externally, and which is tangent to the lines $A_{i-1}A_i, A_iA_{i+1}$, and $A_{i+1}A_{i+2}$ (indices are considered modulo 4 so $A_4 = A_0, A_5 = A_1$, and $A_6 = A_2$). Let T_i be the point of tangency of ω_i with the side A_iA_{i+1}. Prove that the lines A_1A_2, A_3A_4, and T_2T_4 are concurrent if and only if the lines A_2A_3, A_4A_1, and T_1T_3 are concurrent.

Problem 35. Let P be a point inside an arbitrary triangle ABC. Prove that

$$\frac{1}{PA} + \frac{1}{PB} + \frac{1}{PC} \geq \frac{1}{R_a} + \frac{1}{R_b} + \frac{1}{R_c},$$

where R_a, R_b, R_c are the circumradii of triangles PBC, PCA, PAB, respectively.

Problem 36. Through the vertex A of a triangle ABC, a straight line AD is drawn, cutting the side BC at D. Let I be the incenter of triangle ABC, let P be the center of the circle which touches DC, DA, and the circumcircle of ABC internally, and let Q be the center of the circle which touches DB, DA, and the circumcircle of ABC internally. Prove that P, I, Q are collinear.

Problem 37. Let ABC be a triangle with circumcircle Γ and let A' be a point on the side BC. Let $\mathcal{T}_1$ and $\mathcal{T}_2$ be the circles tangent simultaneously to AA', BA', Γ and AA', CA', Γ, respectively. Prove that

(a) If A' is the foot of the internal angle bisector from A, then $\mathcal{T}_1$ and $\mathcal{T}_2$ are tangent to each other at the incenter of triangle ABC.

(b) If A' is the tangency point of the A-excircle with BC, then $\mathcal{T}_1$ and $\mathcal{T}_2$ are congruent.

Problem 38. Suppose M and N are distinct points in the plane of triangle ABC such that $AM : BM : CM = AN : BN : CN$. Prove that the line MN contains the circumcenter of triangle ABC.

Problem 39. The points M, N, P are chosen on the sides BC, CA, AB of a triangle ABC, such that the triangle MNP is acute-angled. We denote with x the length of the shortest altitude of the triangle ABC, and with y the length of the longest altitude of the triangle MNP. Prove that $x \leq 2y$.

Problem 40. Let ABC be a nonisosceles triangle, for which denote by X, Y, Z the tangency points of its incircle with the sides BC, CA and AB, respectively. Let D be the intersection of OI with the sideline BC, where O, I are the circumcenter and incenter, respectively. The line perpendicular to YZ through X cuts AD at E. Prove that the line YZ is the perpendicular bisector of EX.

Problem 41. Six points are chosen on the sides of an equilateral triangle ABC: A_1, A_2 on BC, B_1, B_2 on CA and C_1, C_2 on AB, such that they are the vertices of a convex hexagon $A_1A_2B_1B_2C_1C_2$ with equal side lengths. Prove that the lines A_1B_2, B_1C_2 and C_1A_2 are concurrent.

Problem 42. Given triangle ABC, its centroid G, and its incenter I, construct, using only an unmarked straightedge, its orthocenter H.

Problem 43. Let ABC be a scalene triangle and let Ω be a circle that intersects the side BC at A_1, A_2, the side CA at B_1, B_2, and the side AB at C_1, C_2. Let the tangents at A_1 and A_2 with respect to Ω meet at P, and similarly define R and S. Prove that lines AP, BQ, CR are concurrent.

Problem 44. Let the sides AD and BC of the quadrilateral $ABCD$ (such that AB is not parallel to CD) intersect at point P. Points O_1 and O_2 are circumcenters and points H_1 and H_2 are orthocenters of triangles ABP and CDP, respectively. Denote the midpoints of segments O_1H_1 and O_2H_2 by E_1 and E_2, respectively. Prove that the perpendicular from E_1 on CD, the perpendicular from E_2 on AB and the lines H_1H_2 are concurrent.

Problem 45. Let ABC be an acute-angled triangle with circumcircle $\Gamma(O)$ and let ℓ be a line in the plane which intersects the lines BC, CA, AB at X, Y, Z, respectively. Let ℓ_A, ℓ_B, ℓ_C be the reflections of ℓ across BC, CA, AB, respectively. Furthermore, let M be the Miquel point of triangle ABC with respect to line ℓ.

(a) Prove that lines ℓ_A, ℓ_B, ℓ_C determine a triangle whose incenter lies on the circumcircle of triangle ABC.

(b) If S is the incenter from (a) and O_a, O_b, O_c denote the circumcenters of triangles AYZ, BZX, CXY, respectively, prove that the circumcircles of triangles SOO_a, SOO_b, SOO_c are concurrent at a second point, which lies on Γ.

Problem 46. Let $ABCDEF$ be a convex hexagon, all of whose sides are tangent to a circle ω with center O. Suppose that the circumcircle of triangle ACE is concentric with ω. Let J be the foot of the perpendicular from B to CD. Suppose that the perpendicular from B to DF intersects the line EO at a point K. Let L be the foot of the perpendicular from K to DE. Prove that $DJ = DL$.

Problem 47. A non-isosceles acute triangle ABC is given. Let O, I, H be the circumcenter, the incenter, and the orthocenter of the triangle ABC, respectively. Prove that $\angle OIH > 135°$.

Problem 48. Let ABC be a triangle with circumcenter O and orthocenter H. Parallel lines α, β, γ are drawn through the vertices A, B, C, respectively. Let α', β', γ' be the reflections of α, β, γ in the sides BC, CA, AB, respectively. Prove that these reflections are concurrent if and only if α, β, γ are parallel to the Euler line OH of triangle ABC.

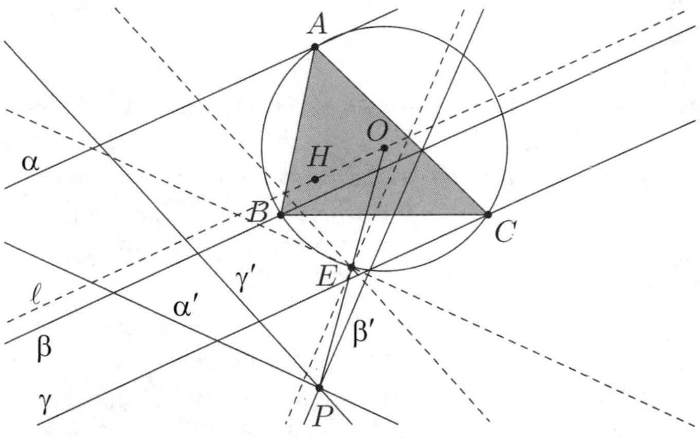

Problem 49. Let ABC be a triangle with circumcircle Γ and nine-point circle γ. Let X be a point on γ, and let Y, Z be on Γ so that the midpoints of segments XY and XZ are on γ.

(a) Prove that the midpoint of YZ is on γ.

(b) Find the locus of the symmedian point of triangle XYZ, as X moves along Γ.

Problem 50. Let ABC be a triangle with circumcircle ω and ℓ a line not intersecting ω. Denote by P the foot of the perpendicular from the center of ω to ℓ. The side-lines BC, CA, AB intersect ℓ at the points X, Y, Z different from P. Prove the circumcircles of triangles AXP, BYP and CZP have a common point different from P or are mutually tangent at P.

Problem 51. Let ABC be an acute-angled triangle and let H be its orthocenter. Let t_a, t_b, t_c be the inradii of triangles HBC, HCA, and HAB, respectively. Prove that

$$t_a + t_b + t_c \leq (6\sqrt{3} - 9)r,$$

where r is the inradius of triangle ABC.

Problem 52. Let ABC be an acute angled triangle satisfying the conditions $AB > BC$ and $AC > BC$. Denote by O and H the circumcenter and orthocenter, respectively, of the triangle ABC. Suppose that the circumcircle of the triangle AHC intersects the line AB at M different from A, and the circumcircle of the triangle AHB intersects the line AC at N different from A. Prove that the circumcenter of the triangle MNH lies on the line OH.

Problem 53. Let $ABCD$ be a rectangle and let ω be an arbitrary circle passing through vertices A and C. Let Γ_1, Γ_2 be the circles inside $ABCD$ so that Γ_1 is tangent to AB, BC, and Γ, and Γ_2 is tangent to CD, DA, and Γ. Prove that the sum of the radii of Γ_1 and Γ_2 is independent of the choice of ω.

Problem 54. Given is an acute-angled triangle ABC and the points A_1, B_1, C_1, which are the feet of its altitudes from A, B, C, respectively. A circle passes through A_1 and B_1 and touches the smaller arc AB of the circumcircle of ABC at point C_2. Points A_2 and B_2 are defined analogously. Prove that the lines A_1A_2, B_1B_2, C_1C_2 have a common point, which lies on the Euler line of triangle ABC.

Problem 55. Let ABC be a triangle with nine-point circle ω. Prove that

(a) There exist precisely three points on the circumcircle of triangle ABC whose Simson lines with respect to the triangle ABC are tangent to ω.

(b) The triangle formed by the three points from (a) is equilateral.

Problem 56. Let ABC be an acute triangle with circumcircle ω. Let ℓ be a tangent line to ω. Let ℓ_a, ℓ_b and ℓ_c be the lines obtained by reflecting ℓ in the lines BC, CA and AB, respectively. Show that the circumcircle of the triangle determined by the lines ℓ_a, ℓ_b and ℓ_c is tangent to the circle ω.

Problem 57. Let M be a point inside triangle ABC with circumcircle (O). Let A_1, B_1, C_1 be the intersections of AM, BM, CM with (O) which are different from the vertices of the triangle. Furthermore, let A_2, B_2, C_2 be the reflections of A_1, B_1, C_1 across the sidelines BC, CA, and AB, respectively. Prove that the triangles $A_1B_1C_1$ and $A_2B_2C_2$ are similar.

Problem 58. Let P be a given point inside the triangle ABC. Suppose L, M, N are the midpoints of BC, CA, AB, respectively, and

$$PL : PM : PN = BC : CA : AB.$$

The extensions of AP, BP, CP meet again the circumcircle of ABC at points D, E, F, respectively. Prove that the circumcenters of triangles PBF, PCE, PCD, PAF, PAE, PBD are concyclic.

Problem 59. Let ABC be an acute-angled triangle and let τ be the inradius of its orthic triangle. Prove that

$$r \geq \sqrt{R\tau},$$

where r and R are the inradius and circumradius of triangle ABC, respectively.

Problem 60. Let ABC be a fixed triangle, and let A_1, B_1, C_1 be the midpoints of sides BC, CA, AB, respectively. Let P be a variable point on the circumcircle. Let lines PA_1, PB_1, PC_1 meet the circumcircle again at A', B', C', respectively. Assume that the points A, B, C, A', B', C' are distinct, and lines AA', BB', CC' form a triangle. Prove that the area of this triangle does not depend on P.

Problem 61. Let P be a point in the plane of a triangle ABC. The lines AP, BP, CP intersect the lines BC, CA, AB at the points A', B', C'. Let Q be the isogonal conjugate of the point P with respect to triangle ABC. Prove that the reflections of the lines AQ, BQ, CQ in the lines $B'C'$, $C'A'$, $A'B'$ are concurrent.

Problem 62. Let triangle ABC be a triangle with circumcircle (O) and incircle (I). Let D, E, F be the tangency points of (I) with BC, CA, AB, respectively. The line EF intersects the circumcircle (O) at X_1, X_2. Similarly, we define Y_1, Y_2 and Z_1, Z_2. Prove that the radical center of the circumcircles of triangles DX_1X_2, EY_1Y_2, FZ_1Z_2 is the orthocenter of triangle DEF.

Problem 63. Let M be an arbitrary point on the circumcircle of triangle ABC and let the tangents from this point to the incircle of the triangle meet the sideline BC at X_1 and X_2. Prove that the second intersection of the circumcircle of triangle MX_1X_2 with the circumcircle of ABC coincides with the tangency point of the circumcircle with the A-mixtilinear incircle.

Problem 64. Let A_1, B_1, C_1 be points on the sides BC, CA, AB of triangle ABC such that the lines AA_1, BB_1, CC_1 are concurrent. Construct three circles Γ_1, Γ_2, and Γ_3 outside the triangle such that they are tangent to the sides of ABC at A_1, B_1, and C_1, respectively, and also tangent to the circumcircle of ABC. Prove that the circle tangent externally to these three circles is also tangent to the incircle of triangle ABC.

Problem 65. Let ABC be a triangle and let P be a point in its plane. Let D, E, F be three points on the lines through P perpendicular to the lines BC, CA, and AB, respectively. Prove that if triangle DEF is equilateral and if P lies on the Euler line of ABC, then the centroid of DEF also lies on the Euler line of ABC.

Problem 66. Let $ABCD$ be a cyclic quadrilateral with circumcircle Γ and let E be an arbitrary point on the side AB. Let DE meet BC at F, DE meet again Γ at P, BP meet AF at Q, and QE meet CD at V. Prove that point V is independent of the position of E.

Problem 67. Let ABC be a triangle and let I and O denote its incenter and circumcenter, respectively. Let ω_A be the circle through B and C that is tangent to the incircle of the triangle ABC; the circles ω_B and ω_C are defined similarly. The circles ω_B and ω_C meet at a point A' distinct from A; the points B' and C' are defined similarly. Prove that the lines AA', BB' and CC' are concurrent at a point on the line IO.

Problem 68. Let ABC be a triangle with circumcircle Ω. Points X and Y are on Ω such that XY meets AB and AC at D and E, respectively. Prove that the midpoints of segments XY, BE, CD, and DE are concyclic.

Problem 69. If two perpendicular straight lines are drawn through the orthocenter of a triangle, they intercept a segment on each of the sidelines. Prove that the midpoints of these three segments are collinear.

Problem 70. Suppose that ABC is an equilateral triangle and that P is a point in the plane of ABC. The perpendicular from P to BC meets AB at X, the perpendicular from P to CA meets BC at Y, and the perpendicular from P to AB meets CA at Z.

(a) If P is in the interior of triangle ABC, prove that the area of XYZ is not greater than the area of ABC.

(b) If P lies on the circumcircle of ABC, prove that X, Y, and Z are collinear.

Problem 71. A triangle ABC is inscribed in a circle ω. A variable line ℓ chosen parallel to BC meets segments AB, AC at points D, E, respectively, and meets ω at points K, L (where D lies between K and E). Circle γ_1 is tangent to the segments KD and BD and also tangent to ω, while circle γ_2 is tangent to the segments LE and CE and also tangent to ω. Determine the locus, as ℓ varies, of the meeting point of the common inner tangents to γ_1 and γ_2.

Problem 72. Let $ABCD$ be a cyclic quadrilateral with circumcircle Γ. Let E be the intersection of AB and CD, F the intersection of AD and BC, and

M, N the midpoints of AC and BD, respectively. Prove that EF is tangent to the circumcircle of triangle MNF.

Problem 73. Consider triangle ABC, and three squares $BCDE$, $CAFG$ and $ABHI$ constructed on its sides, outside the triangle. Let XYZ be the triangle enclosed by the lines EF, DI and GH. Prove that $[XYZ] \le (4-2\sqrt{3})[ABC]$.

Problem 74. A triangle is divided by its three medians into six smaller triangles. Show that the circumcenters of these smaller triangles lie on a circle.

Problem 75. Let H be the orthocenter of an acute triangle ABC with circumcircle Γ. Let P be a point on the arc AB (not containing C) of Γ, and let M be a point on the arc CA (not containing B) of Γ such that H lies on the segment PM. Let K be another point on Γ such that KM is parallel to the Simson line of P with respect to triangle ABC. Let Q be another point on Γ such that $PQ \parallel AB$. Segments AB and KQ intersect at a point J. Prove that triangle KJM is an isosceles triangle.

Problem 76. Let ABC be a triangle and let D, E, F be the tangency points of the incircle with BC, CA, AB, respectively. Let EF meet the circumcircle Γ of ABC at X and Y. Furthermore, let T be the second intersection of the circumcircle of DXY with the incircle. Prove that AT passes through the tangency point A' of the A-mixtilinear incircle with Γ.

Problem 77. Let ABC be a triangle with incircle γ and circumcircle Γ. Let Ω be the circle tangent to the rays AB, AC and to Γ externally, and let A' be the tangency point of Ω with Γ. Furthermore, draw the tangents from A' with respect to incircle γ and let B', C' be their other intersections with Γ. If X denotes the tangency point of the chord $B'C'$ with γ, prove that the circumcircle of triangle BXC is tangent to γ.

Problem 78. In triangle ABC, let D, E, F be the feet of the altitudes from A, B, and C, respectively. Let H be the orthocenter of triangle ABC and let I_1, I_2, I_3 be the incenters of triangles EHF, FHD, and DHE, respectively. Prove that the lines AI_1, BI_2, CI_3 are concurrent.

Problem 79. Let ABC be a triangle with incircle Γ and let D, E, F be the tangency points of Γ with BC, CA, AB, respectively. Let M, N, P be the midpoints of BC, CA, AB, and let X, Y, Z be points on the lines AI, BI, CI, respectively. Prove that the lines XD, YE, ZF are concurrent if and only if the lines XM, YN, ZP are concurrent.

Problem 80. Consider a convex quadrilateral and the incircles of the triangles determined by one of its diagonals. Prove that the tangency points of the incircles with the diagonal are symmetric with respect to the midpoint of the diagonal if and only if the diagonals and the line through the incenters are concurrent.

Problem 81. A circle $\mathcal{K}$ passes through the vertices B, C of triangle ABC and another circle ω touches AB, AC, $\mathcal{K}$ at P, Q, T, respectively. If M is the midpoint of the arc BC of $\mathcal{K}$ that contains the point T, prove that the lines BC, PQ, and MT are concurrent.

Problem 82. Let $ABCD$ be a quadrilateral with an inscribed circle Γ and let γ be an arbitrary line tangent to its incircle. Let A', B', C', D' be the orthogonal projections of A, B, C, D onto γ, respectively. Prove that

$$\frac{AA' \cdot CC'}{BB' \cdot DD'} = \frac{AO \cdot CO}{BO \cdot DO}.$$

Problem 83. Let $ABCD$ be a quadrilateral with an inscribed circle and let P be the intersection of its diagonals. Prove that the incenters of triangles PAB, PBC, PCD, PDA are concyclic.

Problem 84. Let P be an arbitrary interior point of an equilateral triangle ABC. Prove that

$$|\angle PAB - \angle PAC| \geq |\angle PBC - \angle PCB|.$$

Problem 85. Let ρ be the inscribed circle of ABC with incenter I and let D, E and F be the points of tangency of ρ with BC, CA, and AB respectively. Denote by M the second intersection of ρ with AD, and let N be the second intersection of the circumcircle of triangle CDM with DF. Furthermore, let G be the intersection of CN and AB. Prove that $CD = 3FG$.

Problem 86. Let P be a point inside triangle ABC such that

$$\angle PAB + \angle PBC + \angle PCA = 90^\circ.$$

If P' is the isogonal conjugate of P with respect to triangle ABC, prove that PP' passes through the circumcenter of triangle ABC.

Problem 87. Let the tangents to the circumcircle of a triangle ABC at the vertices B and C intersect each other at a point X. Consider the circle $\mathcal{X}$ centered at X with radius XB, and let M be the point of intersection of the internal angle bisector of angle A with $\mathcal{X}$ such that M lies in the interior of triangle ABC. If O is the circumcenter of ABC, denote by P the intersection of OM with the sideline BC, and let E, F be the orthogonal projections of M on CA, AB, respectively. Prove that the lines PE and FP are perpendicular.

Problem 88. Let ABC be a triangle with orthocenter H and let P be the second intersection of the circumcircle of triangle AHC with the internal bisector of the angle $\angle BAC$. Let X be the circumcenter of triangle APB and Y the orthocenter of triangle APC. Prove that the length of segment XY is equal to the circumradius of triangle ABC.

Problem 89. Let C_1, C_2 and C_3 be circles that do not intersect and so that none of them is inside another. Let (L_1, L_2), (L_3, L_4) and (L_5, L_6) be internal common tangents of the pairs of circles (C_1, C_2), (C_1, C_3), (C_2, C_3), respectively. Furthermore, let L_1, L_2, L_3, L_4, L_5, L_6 bound a hexagon $AC'BA'CB'$, where the vertices are labeled in cyclic order. Prove that the lines AA', BB', CC' are concurrent.

Problem 90. Let ABC be a triangle, let E, F be the tangency points of the incircle $\Gamma(I)$ to the sides AC, respectively AB, and let M be the midpoint of the side BC. Let $N = AM \cap EF$, let γ be the circle of diameter BC, and let X, Y be the other (than B, C) intersection points of BI, respectively CI, with γ. Prove that

$$\frac{NX}{NY} = \frac{AC}{AB}.$$

Problem 91. Let the excircle of the triangle ABC lying opposite to A touch its side BC at the point A_1. Define the points B_1 and C_1 analogously. Suppose that the circumcenter of triangle $A_1B_1C_1$ lies on the circumcircle of triangle ABC. Prove that triangle ABC is right-angled.

Problem 92. Let ABC be a triangle with excenters I_a, I_b, I_c and A-excircle Γ_a. Let ℓ_1, ℓ_2 be the common tangents of Γ_a and the circumcircle Ω of triangle ABC, and suppose that these meet the line BC at points D and E, respectively. Furthermore, let P and Q be the tangency points of Ω with ℓ_1 and ℓ_2, respectively. Prove that P, Q, I_b, I_c are concyclic.

Problem 93. Let ABC be a triangle with circumcircle Γ and excircles Γ_a, Γ_b, Γ_c. Let T_A be the intersection of the tangents at B and C with respect to Γ. Define the points T_B and T_C similarly and let D be an arbitrary point on the side T_BT_C. Furthermore, let the tangent from D to Γ_c, which is different from the line T_BT_C, intersect the line T_CT_A at E, and let the tangent from D to Γ_b intersect T_AT_B at F. Prove that the line EF is tangent to Γ_a.

Problem 94. The point M is inside the convex quadrilateral $ABCD$, such that $MA = MC$, $\angle AMB = \angle MAD + \angle MCD$, and $\angle CMD = \angle MCB + \angle MAB$. Prove that $AB \cdot CM = BC \cdot MD$ and $BM \cdot AD = MA \cdot CD$.

Problem 95. The circle passing through the feet of the internal angle bisectors of a triangle contains the Feuerbach point of the triangle.

Problem 96. Let $ABCD$ be a cyclic quadrilateral. If there exists a point X such that the angles $\angle XAD$, $\angle XBA$, $\angle XCB$, $\angle XDC$ are all equal, prove that $ABCD$ is a harmonic quadrilateral.

Problem 97. Let ABC be an isosceles triangle with $AB = AC$ and let M be the midpoint of BC. Find the locus of points P inside ABC for which $\angle BPM + \angle CPA = 180°$.

Problem 98. Let ABC be a triangle and let M be the midpoint of BC and X be the midpoint of the A-altitude of ABC. Prove that the symmedian point of ABC lies on the line MX.

Problem 99. Let ABC be a triangle. Its excircles touch sides BC, CA, AB at D, E, F, respectively. Prove that the perimeter of triangle ABC is at most twice that of triangle DEF.

Problem 100. Let ABC and $A'B'C'$ be two triangles such that the perpendiculars from vertices A, B, C to the sides $B'C'$, $C'A'$, $A'B'$ are concurrent at some point O. Prove that:

(a) Perpendicular lines from vertices A', B', C' to the sides BC, CA, AB are concurrent at some point O'.

(b) If $O = O'$, then lines AA', BB', CC' are concurrent.

(c) If $O \neq O'$, but the lines AA', BB', CC' are still concurrent at some point P, then line OO' passes through P and is perpendicular to the perspectrix of triangles ABC and $A'B'C'$.

Problem 101. Let ABC be a non-equilateral triangle with circumcircle Γ and let A', B', C' be the reflections of A, B, C across the sides BC, CA, AB, respectively. Let A_t, B_t, C_t be the pairwise intersections of the tangents at the vertices of ABC with respect to Γ. Prove that the circumcircles of triangles $A_tB'C'$, $B_tC'A'$, $C_tA'B'$ are concurrent.

Problem 102. Four segments divide a convex quadrilateral into nine quadrilaterals as in the diagram. The points of intersections of these segments lie on the diagonals of the quadrilateral. It is known that quadrilaterals 1, 2, 3, 4 admit inscribed circles. Prove that the quadrilateral 5 also has an inscribed circle.

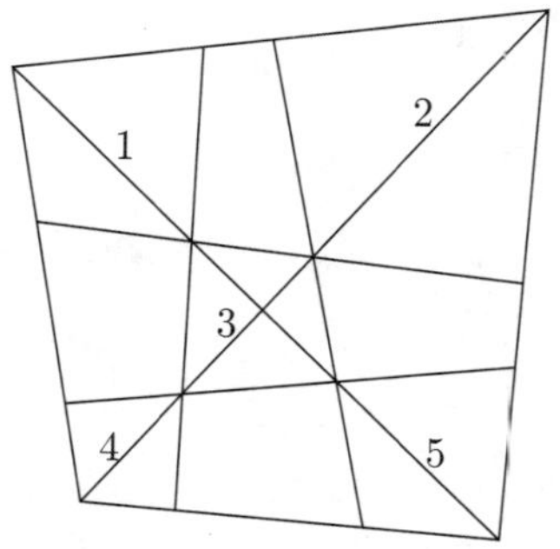

Problem 103. Given a triangle ABC, let C_A be the circle tangent to the sides AB, AC and to the circumcircle internally. Define C_B and C_C analogously. Find the triangle, unique up to similarity, for which the inradius and the radii of the three circles C_A, C_B and C_C are in arithmetic progression.

Problem 104. Let $BCKL$, $CAHF$, $ABDE$ be rectangles constructed on the sides BC, CA, AB of triangle ABC. Let A' be the intersection of EK with HL, B' the intersection of HL with DF, and C' the intersection of DF with EK. Prove that the lines AA', BB', CC' are concurrent.

Problem 105. Let Γ be a circle in plane and let γ_A, γ_B, γ_C, γ_D, γ_E, and γ_F be six circle inside it, each internally tangent to Γ and tangent to each other in pairs. Let A, B, C, D, E, and F be the consecutive points of tangency of the small circles with the outer circle. Prove that the lines AD, BE, and CF are concurrent.

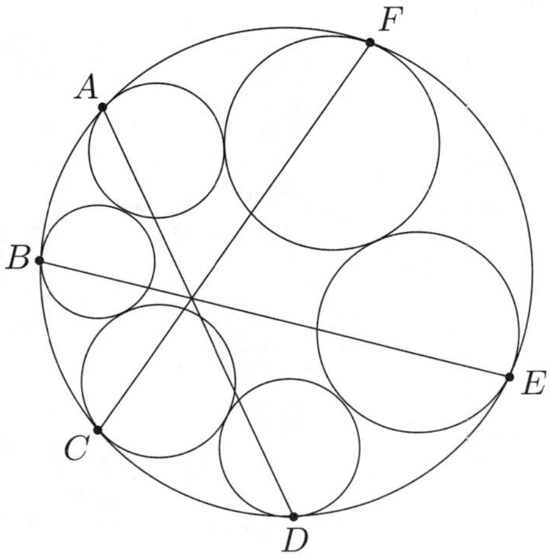

Problem 106. Let $ABCD$ be a convex quadrilateral and let r be a real number. Find the locus of the points P in plane for which the sum of areas $[PAB] + [PCD]$ equals r is a line.

Problem 107. Let $ABCD$ be a convex quadrilateral and let M, N be the midpoints of diagonals AC and BD, respectively. Let $XYZT$ be the quadrilateral determined by the internal angle bisectors of its four angles. Prove that $XYZT$ is cyclic and that points M, N and the anticenter of $XYZT$ are collinear.

Problem 108. Let ABC be a triangle. Find all points P on segment BC with the following property: If X, Y are the intersections of line PA with the common external tangents of the circumcircles of triangles PAB and PAC, then

$$\left(\frac{PA}{XY}\right)^2 + \frac{PB \cdot PC}{AB \cdot AC} = 1.$$

Problem 109. Let ABC be a triangle with side lengths a, b, c, and area S and let x, y, z be three positive real numbers. Then,

$$a^2+b^2+c^2 \geq 4\sqrt{3}S+\frac{2}{x+y+z}\left(\frac{x^2-yz}{x}\cdot a^2+\frac{y^2-zx}{y}\cdot b^2+\frac{z^2-xy}{z}\cdot c^2\right).$$

Problem 110. Let ABC be a triangle and let D_1 ,D_2 be the tangency points of the $A-$excircle with the lines AB and AC, respectively. Similarly, define B_1, B_1, and C_1, C_2. Furthermore, let E_1E_2 and F_1F_1 meet at X, F_1F_2 and D_1D_2 meet at Y, and D_1D_2 and E_1E_2 meet at Z. Prove that the points X, Y, Z lie on the altitudes of triangle ABC.

Solutions

Problem 1. (Haruki's Lemma) Given two nonintersecting chords AB and CD in a circle and a variable point P on the arc AB remote from points C and D, let E and F be the intersections of chords PC, AB, and of PD, AB, respectively. Prove that the value of

$$\frac{AE \cdot BF}{EF}$$

does not depend on the position of P.

Solution. The proof relies on the fact that the angle $\angle CPD$ is constant.

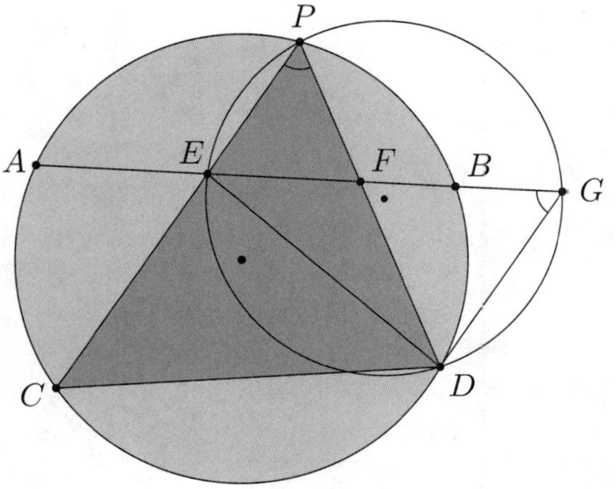

We begin by constructing the circumcircle of triangle PED. Define point G to be the intersection of this circle with the line AB. Note that $\angle EGD = \angle EPD$ as they are subtended by the same chord ED of the circumcircle of triangle PED; these angles remain constant as P varies on the arc AB. Hence, for all positions of P, $\angle EGD$ remains fixed and, therefore, point G remains fixed on the line AB. It follows that BG is constant.

On the other hand, by Power of Point, we have that $AF \cdot FB = PF \cdot FD$ and $EF \cdot FG = PF \cdot FD$. Hence,

$$(AE + EF) \cdot FB = EF \cdot (FB + BG),$$

and $AE \cdot FB = EF \cdot BG$. Therefore, we conclude that

$$\frac{AE \cdot BF}{EF} = BG,$$

a constant.

■

Comment. Haruki's lemma gives a very short proof of the celebrated Butterfly Theorem.

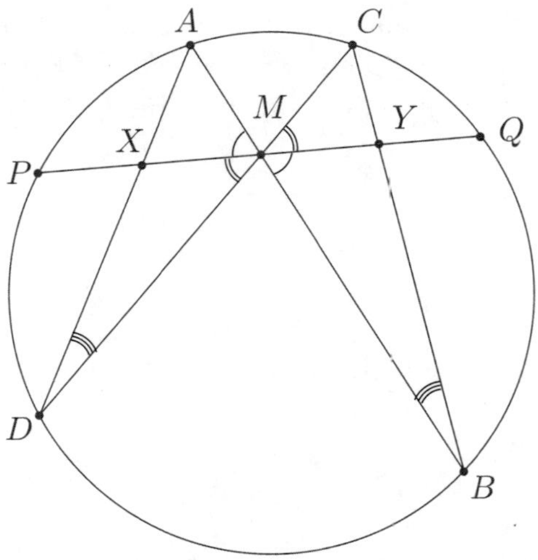

Butterfly Theorem. Let M be the midpoint of chord PQ of a given circle, through which two other chords AB and CD are drawn; AD cuts PQ at X and BC cuts PQ at Y. Then, M is also the midpoint of XY.

Proof. We think of A and C as being two positions of the variable point traversing the circle. Then, Haruki's lemma tells us that

$$\frac{XP \cdot MQ}{XM} = \frac{MP \cdot YQ}{YM},$$

which, because of $MP = MQ$, is simplified to

$$\frac{XP}{XM} = \frac{YQ}{YM}.$$

Adding 1 to both sides gives

$$\frac{XP + XM}{XM} = \frac{YQ + YM}{YM}.$$

Applying $MP = MQ$ again, we obtain the required $XM = YM$. This completes the proof.

Problem 2. (Nikolaos Dergiades, Forum Geometricorum) Consider a triangle ABC with $a \leq b \leq c$. Denote by X, Y, Z the midpoints of the sides BC, CA, and AB, respectively. Let D, E, F be points on the sides BC, CA, AB satisfying the following two conditions:

(i) D is between X and C, E is between Y and C, and F is between Z and B.

(ii) $\angle CDE \leq \angle BDF$, $\angle CED \leq \angle AEF$, and $\angle BFD \leq \angle AFE$.

Prove that the perimeter of triangle DEF does not exceed the semiperimeter of triangle ABC.

Solution. Denote by $\mathbf{i},\mathbf{j},\mathbf{k}$ the unit vectors along $\overrightarrow{EF}$, $\overrightarrow{FD}$, $\overrightarrow{DE}$, respectively. Since $\angle BFD \le \angle AFE$, we have $\mathbf{i}\cdot\overrightarrow{ZF} \le \mathbf{j}\cdot\overrightarrow{ZF}$. Similarly, since $\angle CDE \le \angle BDF$ and $\angle CED \le \angle AEF$, we have

$$\mathbf{j}\cdot\overrightarrow{XD} \le \mathbf{k}\cdot\overrightarrow{XD} \quad\text{and}\quad \mathbf{i}\cdot\overrightarrow{EY} \le \mathbf{k}\cdot\overrightarrow{EY}.$$

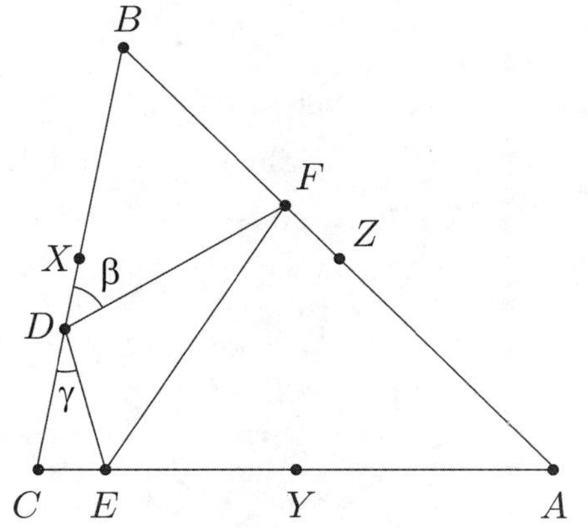

Now, we have

$$\begin{aligned}
EF+FD+DE &= \mathbf{i}\cdot\overrightarrow{EF}+\mathbf{j}\cdot\overrightarrow{FD}+\mathbf{k}\cdot\overrightarrow{DE}\\
&= \mathbf{i}\cdot(\overrightarrow{EY}+\overrightarrow{YZ}+\overrightarrow{ZF})+\mathbf{j}\cdot(\overrightarrow{FZ}+\overrightarrow{ZX}+\overrightarrow{XD})+\mathbf{k}\cdot\overrightarrow{DE}\\
&\le (\mathbf{k}\cdot\overrightarrow{EY}+\mathbf{i}\cdot\overrightarrow{YZ}+\mathbf{j}\cdot\overrightarrow{ZF})+(\mathbf{j}\cdot\overrightarrow{FZ}+\mathbf{j}\cdot\overrightarrow{ZX}+\mathbf{k}\cdot\overrightarrow{XD})+\mathbf{k}\cdot\overrightarrow{DE}\\
&= \mathbf{i}\overrightarrow{YZ}+\mathbf{j}\overrightarrow{ZX}+\mathbf{k}\overrightarrow{XY}\\
&\le |\mathbf{i}|\cdot|\overrightarrow{YZ}|+|\mathbf{j}|\cdot|\overrightarrow{ZX}|+|\mathbf{k}|\cdot|\overrightarrow{XY}|\\
&= YZ+ZX+XY\\
&= \frac{1}{2}(AB+BC+CA).
\end{aligned}$$

Equality holds in the above only when the triangles DEF and XYZ have parallel sides, i.e., when the points D, E, F coincide with the midpoints X, Y, Z, respectively, as is easily seen.

■

Problem 3. (Virgil Nicula and Cosmin Pohoata, Journal of Geometry and Graphics) Let A' be the foot of the internal angle bisector of the angle BAC of a given triangle ABC. Let P be an arbitrary point on the cevian AA', different from A', and denote by B', C' the intersections of the lines BP, CP, with the sidelines CA, and AB, respectively. If $BB' = CC'$, prove that $AB = AC$.

Solution. We start with a simple preliminary result that will be used in the proof of the main theorem.

Claim. In triangle ABC, let the two cevians BB' and CC' intersect at P. If $BB' = CC'$, then $PB' < PC$ and $PC' < PB$.

Suppose that $PB' \geq PC$. Since $BB' = CC'$, it follows that $PC' \geq PB$. Therefore

$$\begin{aligned}\angle B'CP &\geq \angle CB'P, \quad \text{because } PB' \geq PC\\ &> \angle ABB', \quad \text{by the exterior angle theorem}\\ &\geq \angle PC'B, \quad \text{because } PC' \geq PB\\ &> \angle B'CP, \quad \text{by the exterior angle theorem.}\end{aligned}$$

Thus we reach the contradiction $\angle B'CP > \angle B'CP$. Therefore, $PB' < PC$ and similarly, $PC' < PB$. This proves the Claim.

The main result will be now split in two parts.

Part 1. Let A' be the foot of the internal angle bisector of the angle BAC of a given triangle ABC. Consider an arbitrary point P on the ray AA', different from A', and denote by B', C' the intersections of the lines BP, CP with the sidelines CA and AB, respectively. If $BB' = CC'$, then $AB = AC$.

Proof. Erect a triangle $C'XC$ on the segment CC' that is congruent with triangle BAB' and such that the points B and X do not lie on the same side of AC.

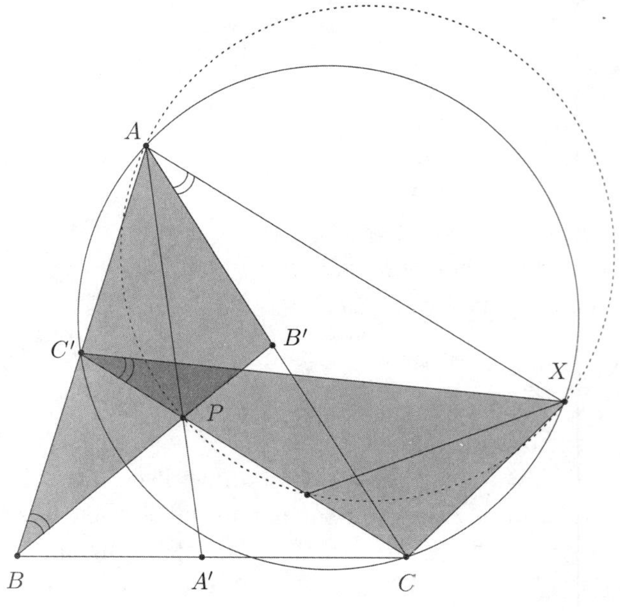

We have that the the angles $\angle C'AC$ and $\angle C'XC$ are equal, and thus the quadrilateral $C'AXC$ is cyclic, which means that $\angle CAX = \angle CC'X$. On the other hand, the angles $\angle CC'X$ and $\angle B'BA$ are equal, and therefore, $\angle CAX = \angle B'BA$.

Let P' be the foot of the internal angle bisector of the angle $C'XC$ in triangle $C'XC$. Since triangles $C'XC$ and BAB' are congruent, the previous Claim yields $CP' = B'P < CP$, which means that P' lies between C and C'. Moreover,

$$\angle CP'X = \angle B'PA = \angle BAP + \angle B'BA = \angle PAC + \angle CAX = \angle PAX.$$

From this we deduce that the quadrilateral $AXP'P$ is cyclic, and plus, since the segments AP and XP' are congruent, the quadrilateral $AXP'P$ is an isosceles trapezoid, and thus, we conclude that the lines AX and CC' are parallel. It now follows that $\angle CAX = \angle ACC'$, and hence, $\angle B'BA = \angle ACC'$. From this and the assumption $BB' = CC'$ we conclude that the triangles ABB' and ACC' are congruent, and therefore $AB = AC$. This proves Part 1.

Part 2. Let A' be the foot of the internal angle bisector of the angle BAC of a given triangle ABC. Consider a point P on the ray AA' beyond A', and denote by B', C' the intersections of the lines BP, CP, with the sidelines CA and AB, respectively. If $BB' = CC'$, then $AB = AC$.

Proof. Let A'' be the intersection of AA' with $B'C'$. It follows from Part 1 (applied in triangle $AC'B'$) that $AC' = AB'$. It also follows that A'' is the midpoint of $B'C'$. By Ceva's theorem, we obtain

$$\frac{AB}{BC'} = \frac{AC}{CB'}$$

and therefore $BC \| C'B'$. Thus $AB = AC$, which proves Part 2 and thus the problem.

■

Problem 4. (Taylor's Theorem) Let ABC be a triangle and let A', B', C' be the feet of the altitudes from A, B, C, respectively. Let A_1 be the foot of the perpendicular from A' to AB and let A_2 be the foot of the perpendicular from A' to AC. Furthermore, let Ω_A be the circle centered at vertex A having radius AA'. Analogously, define points B_1, B_2, C_1, C_2, and circles Ω_B and Ω_C. Prove that:

(a) Points A_1, A_2, B_1, B_2, C_1, C_2 are all concyclic.

(b) The center of the circle from (a) has equal powers with respect to Ω_A, Ω_B and Ω_C.

Solution. (a) Notice that quadrilaterals $BCB'C'$ and $C'B_2C_1B'$ are cyclic, so

$$\angle AC_1B_2 = \angle AC'B' = \angle ACB,$$

which means that the lines B_2C_1 and BC are parallel. Similarly, we get that $A_2B_1 \| AB$ and $A_1C_2 \| AC$.

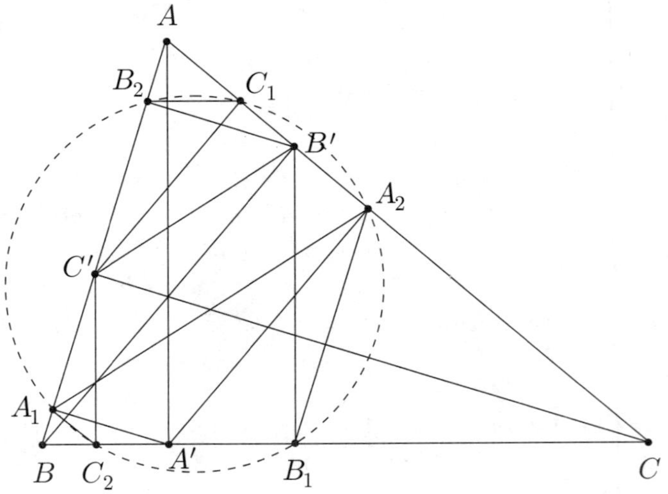

On the other hand, quadrilateral $AA_1A'A_2$ is also cyclic, so

$$\angle AA_1A_2 = \angle AA'A_2 = \angle C = \angle AC_1B_2,$$

so quadrilatral $A_1B_2C_1A_2$ is cyclic. Analogously, we also get that quadrilaterals $C_1A_2B_1C_2$ and $B_1C_2A_1B_2$ are cyclic. Denote their circumcircles by Γ_1, Γ_2, Γ_3, respectively. We want to show that $\Gamma_1 = \Gamma_2 = \Gamma_3$, and we will do this by contradiction.

Notice that if any two of the circles coincide, then we are done, all six points A_1, A_2, B_1, B_2, C_1, C_2 are then concyclic. So, let us suppose that $\Gamma_1 \neq \Gamma_2 \neq \Gamma_3 \neq \Gamma_1$. In this case, they pairwise have radical axes, which must be concurrent at the radical center of the three circles. However, this is impossible, since the three radical axes are in fact the lines BC, CA, AB, which are obviously not concurrent. This proves that we certainly need $\Gamma_i = \Gamma_j$ for some $i \neq j$ in $\{1, 2, 3\}$, which is exactly what we wanted.

Comment. The circle passing through the points A_1, A_2, B_1, B_2, C_1, C_2 is known in literature as the *Taylor circle* of triangle ABC.

(b) Let h_a, h_b, and h_c be the lengths of segments AA', BB', and CC', respectively. We want to prove that

$$AT^2 - h_a^2 = BT^2 - h_b^2 = CT^2 - h_c^2.$$

We begin with a simple observation.

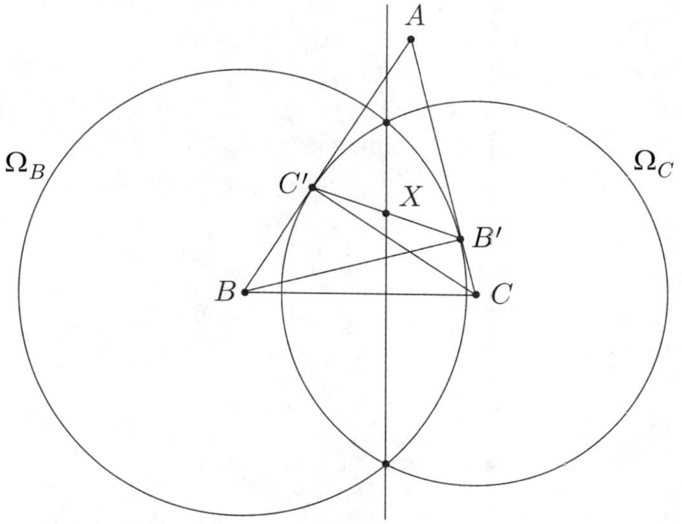

Claim. The midpoint X of segment $B'C'$ lies on the radical axis of circles Ω_B and Ω_C.

Proof. Since BX and CX are medians in triangles $BB'C'$, $CC'B'$, respectively,

$$\begin{aligned} BX^2 - h_b^2 &= \frac{1}{4}\left(2BC'^2 - 2h_b^2 - B'C'^2\right) \\ &= \frac{1}{4}\left(2a^2 - 2h_c^2 - 2h_b^2 - B'C'^2\right) \\ &= \frac{1}{4}\left(2B'C^2 - 2h_c^2 - B'C'^2\right) \\ &= CX^2 - h_c^2. \end{aligned}$$

Hence, X lies on their radical axis.

On other hand, the perpendicular ℓ_1 through X to the side BC is the perpendicular bisector of segment A_1A_2, since X is the midpoint of $B'C'$. Hence, ℓ_1, which in fact is the radical axis of Ω_B and Ω_C, passes through the Taylor center T. Analogously, the perpendiculars ℓ_2, ℓ_3, through the midpoints Y, Z of the segments $C'A'$, $A'B'$ to the sides CA, and AB, respectively, pass through the Taylor center T. Hence, the radical center of the circles centered at A, B, and C with radii h_a, h_b, h_c coincides with the Taylor center of ABC. ■

Problem 5. (Virgil Nicula, AOPS Forum) Let $\mathcal{C}$ be a circle and let P be a point in the exterior of $\mathcal{C}$. The tangents from P intersect the circle at A and B, respectively. Let M be the midpoint of segment AP and N the second intersection of the line BM with circle $\mathcal{C}$. Prove that $PN = 2MN$.

Solution 1. Let X be the second intersection of PN with $\mathcal{C}$. The tangents at A and B meet at P, thus quadrilateral $AXBN$ is harmonic. In particular, this implies that pencil (BA, BX, BB, BN) is harmonic, so by intersecting it with PA, it follows that $BX \| PA$.

Let Y be the reflection of N across M. From the construction, in order to show that $PN = 2MN$, it is sufficient to see that $PN = YN$, or in other words that $\angle YPN = \angle PYN$. But this is now easy: from the parallelisms, note that

$$\begin{aligned} \angle YPN &= \angle YPA + \angle APN \\ &= \angle PAN + \angle NXB \\ &= \angle AXN + \angle NXB \\ &= \angle AXB, \end{aligned}$$

which finishes the argument, since $\angle PYN = \angle YNA = \angle AXB$. This completes the proof.

Solution 2. The second solution is perhaps more like what someone would have found under the pressure of a contest. By Power of Point, $MN \cdot MB = MA^2 = MP^2$. Hence, the line MP is tangent to the circumcircle of triangle PNB, or, in other words, $\angle MPN = \angle PBN$. By the Law of Sines, it follows that

$$\begin{aligned} \frac{PN}{MN} &= \frac{PN}{\sin \angle PBN} \cdot \frac{\sin \angle PBN}{MN} \\ &= \frac{PB}{\sin \angle PNB} \cdot \frac{\sin \angle MPN}{MN} \\ &= \frac{PB}{\sin \angle PNB} \cdot \frac{\sin \angle PNM}{PM} \\ &= \frac{PB}{PM} \\ &= \frac{PA}{PM} \\ &= 2. \end{aligned}$$

This proves that $PN = 2MN$, as claimed.

■

Problem 6. (Virgil Nicula, Danube Mathematical Cup 2010) Let ℓ be a line in the exterior of a given circle $\rho(O)$. Let A be the foot of the perpendicular from O to ℓ, and let M be an arbitrary point on ρ. Furthermore, let X and Y be the second intersections of the circle with diameter AM with ρ and ℓ, respectively. Prove that the line XY passes through a point which is independent of the position of M.

Solution. Let γ be the tangent at A with respect to the circle of diameter AM and let $F = \gamma \cap XM$, $K = AM \cap XY$, and $L = OA \cap XY$. Since $\angle FMA = \angle LYA$ and $\angle YAL = \angle FAM = 90°$, triangles LAY and FAM are similar; thus $\angle ALY = \angle AFM$, which implies that quadrilateral $AFXL$ is cyclic.

On other hand, $AL \perp FL$, hence $FL \perp AL$, i.e., $FL \perp AO$, which immediately implies that $FL \| \ell$. Now, because γ is the radical axis of the circle of diameter AM and the degenerate circle A, and XM is the radical axis of the circle of diameter AM and ρ, the intersection $F = \gamma \cap XM$ is the radical center of the three circles; thus, in particular, F is on the radical axis of ρ and the degenerated circle A. This yields that F is a fixed point. In conclusion, since $FL \perp OA$, it also follows that L is fixed, and since L is on XY, we deduce that L is the desired fixed point. This completes the proof.

■

Problem 7. (AOPS Forum) Let X, Y, Z be the midpoints of the arcs BC, CA, AB, respectively, of triangle ABC containing the vertices of the triangle. Prove that the Simson lines of X, Y, Z with respect to ABC are concurrent.

Solution. Let M, N, P be the midpoints of the sides. We prove that s_X is the M-internal angle bisector of triangle MNP. To see this, we look at the angle $\angle(s_X, MN)$, and note that

$$\angle(s_X, MN) = \angle s_X, AB = \angle A - \angle s_X, AC = \angle A - \angle MXC = \frac{1}{2}\angle A.$$

Hence, s_X bisects $\angle NMP$ (as the angle at M in triangle MNP is equal to $\angle A$). This proves that the Simson lines s_X, s_Y, s_Z are concurrent at the incenter of triangle MNP. This completes the proof.

■

Problem 8. (Mario Dalcin and Eckart Schmidt, Forum Geometricorum) Let ABC be a triangle with circumcenter O and let D, E, F be any three points lying on the sides BC, CA, AB, respectively. Let D', E', F' be the reflections of D, E, F with respect to the midpoints of BC, CA, AB, respectively. Prove that

(a) The Miquel points M of D, E, F and M' of D', E', F' are equidistant from the circumcenter of ABC.

(b) The centroids of triangles DEF and $D'E'F'$ are symmetric with respect to the centroid of ABC.

(c) Triangles DEF and $D'E'F'$ have the same area.

Solution. (a) Let N be the reflection of M about the perpendicular bisector of BC. Let ω be the circle with center O and radius OM and let N' be the point where $D'N$ intersects ω again. Then $\angle N'D'C = \angle MDB$ and $\angle N'OM = 2\angle N'NM = 2\angle MDB$. This means if we define N' with side AB or AC, we will get the same point. We also have $\angle N'D'C = \angle N'E'A = \angle N'F'B$, so N' is the Miquel point of D', E', F', and $ON' = OM$. This proves (a).

(b) Suppose points D, E, F are dividing the sides BC, CA, AB in the ratios

$$BD : DC = x : 1-x, \quad CE : EA = y : 1-y, \quad AF : FB = z : 1-z.$$

In terms of barycentric coordinates with respect to ABC, we have

$$D = (1-x)B + xC, \quad E = (1-y)C + yA, \quad F = (1-z)A + zB.$$

Consequently, by definition, points D', E', F' satisfy

$$D' = xB + (1-x)C, \quad E' = yC + (1-y)A, \quad F' = zA + (1-z)B,$$

and so the centroids G_{DEF} and $G_{D'E'F'}$ of triangles DEF and $D'E'F'$, respectively, satisfy

$$\begin{aligned} G_{DEF} &= \frac{1}{3}(D+E+F) \\ &= \frac{(1+y-z)A + (1+z-x)B + (1+x-y)C}{3} \\ &= \frac{2}{3}(A+B+C) - \frac{(1-y+z)A + (1-z+x)B + (1-x+y)C}{3} \\ &= 2G - G_{D'E'F'}, \end{aligned}$$

where G represents the centroid of ABC. This proves that G_{DEF} and $G_{D'E'F'}$ are symmetric with respect to G, as claimed. This proves (b).

(c) We can prove this part by employing the computations we did in part (b). Indeed, given the barycentric coordinates of points D, E, F, D', E', F', we can write immediately

$$\begin{aligned} [DEF] &= (1 - (x+y+z) + (xy+yz+zx)) \\ &= (xyz + (1-x)(1-y)(1-z)) \\ &= [D'E'F']. \end{aligned}$$

Nonetheless, we shall present a beautiful synthetic approach that only uses an interplay between area additions and subtractions. Note that

$$[CD'E'] = [AD'E] = [AD'C] - [CD'E].$$

Analogously, $[AE'F'] = [BE'A] - [AE'F]$ and $[BF'D'] = [CF'B] - [BF'D]$. Adding these three together, we get

$$\begin{aligned}&[CD'E'] + [AE'F'] + [BF'D'] \\ &= [AD'C] + [BE'A] + [CF'B] - [CD'E] - [AE'F] - [BF'D]. \qquad (1)\end{aligned}$$

Furthermore,

$$[CDE] = [BD'E] = [BEC] - [CD'E].$$

and similarly $[AEF] = [CFA] - [AE'F]$ and $[BFD] = [ADB] - [BF'D]$. Therefore,

$$\begin{aligned}&[CDE] + [AEF] + [BFD] \\ &= [BEC] + [CFA] + [ADB] - [CD'E] - [AE'F] - [BF'D]. \qquad (2)\end{aligned}$$

But $D'C = DB$, $E'A = EC$, $F'B = FA$, so $[AD'C] = [ADB]$, $[BE'A] = [BEC]$, $[CF'B] = [CFA]$. This, together with (1) and (2), yields

$$[CD'E'] + [AE'F'] + [BF'D'] = [CDE] + [AEF] + [BFD]$$

and so $[ABC] - [D'E'F'] = [ABC] - [DEF]$, i.e., $[DEF] = [D'E'F']$, which proves (c).

■

Problem 9. (Cosmin Pohoata, Romania IMO TST 2008) Let ABC be a triangle with $\angle BAC < \angle ACB$. Let D, E be points on the sides AC and AB, respectively, such that the angles ACB and BED are congruent. If F lies in the interior of the quadrilateral $BCDE$ such that the circumcircle of triangle BCF is tangent to the circumcircle of DEF and the circumcircle of BEF is tangent to the circumcircle of CDF, prove that the points A, C, E, F are concyclic.

Solution. Invert around F (with arbitrary power), and write P' for the image of P under this inversion. Now, since the circumcircles of triangles FBE and DFC are tangent, we see that $B'E' \parallel D'C'$. Similarly, $E'D' \parallel B'C'$, so $E'D'C'B'$ is a parallelogram. This means that $\angle B'E'D' = \angle D'C'B'$. Also, the inversion gives

$$\angle BED = \angle FEB + \angle FED = \angle FD'E' + \angle FB'E'.$$

Similarly,

$$\angle ACB = \angle DCB = \angle FCD + \angle FCB = \angle FB'C' + \angle FD'C'$$

so

$$\angle FD'E' + \angle FB'E' = \angle BED = \angle ACB = \angle FB'C' + \angle FD'C'.$$

Also,

$$\begin{aligned}\angle B'FD' &= 360^\circ - (\angle B'E'D' + \angle FB'E' + \angle FD'E') \\ &= 360^\circ - (\angle B'C'D' + \angle FB'C' + \angle FD'C') \\ &= 360^\circ - \angle B'FD',\end{aligned}$$

which implies that $\angle B'FD' = 180^\circ$, so B', F, and D' are collinear. Now, notice that A is the intersection between BE and CD, so A' is the intersection between the circumcircles of $FE'B'$ and $FD'C'$. Moreover,

$$\angle FA'E' = 180^\circ - \angle FB'E' = 180 - \angle FD'C' = 180 - \angle FA'C'$$

so $\angle FA'C' + \angle FA'E' = 180$, which means that E', A', and C' are collinear, and thus $FCAE$ is cyclic.

■

Problem 10. (Dinu Serbanescu, Romania IMO TST 2012) Let γ be a circle and l a line in its plane. Let K be a point on l, located outside of γ. Let KA and KB be the tangents from K to γ, where A and B are distinct points on γ. Let P and Q be two points on γ. Lines PA and PB intersect line l in two points R and respectively S. Lines QR and QS intersect the second time circle γ in points C and D. Prove that the tangents from C and D to γ are concurrent on line l.

Solution. First, note that Pascal's theorem applied to $DAPBCQ$ yields $DA \cap BC \in \ell$. Furthermore, Pascal's theorem for $DBBCAA$ gives the collinearity of $DB \cap CA$, K and $BC \cap AD$, thus $DB \cap CA$ lies on ℓ from the first consequence of Pascal's. Finally, by Pascal's theorem once again, this time applied to $DDACCB$, yields $CC \cap DD \in \ell$, as ℓ is the line determined by $DA \cap CB$ and $AC \cap BD$. This completes the proof.

■

Comment. The following exercise proposed in Mathematical Reflections proves something slightly more general:

Claim. Let Γ be a circle and ℓ be a line lying outside Γ. Let $K \in \ell$ and let AB and CD be chords of Γ passing through K. Let P and Q lie on Γ. Let PA, PB, PC, PD meet ℓ at X, Y, Z, T, respectively, and then let QX, QY, QZ, QT meet again Γ at R, S, U, V, respectively. Prove that RS and UV meet on ℓ.

As before, the trick is to apply Pascal's theorem several times and to relate the collinearities obtained. Consider the hexagon $PAUQRC$. Since $PA \cap QR = X \in \ell$ and $PC \cap QU = Z \in \ell$, by Pascal's theorem, it follows that $L = AU \cap CR \in \ell$. Consider next the hexagon $PBVQSD$. Since $PB \cap QS = Y \in \ell$ and $PD \cap QV = T \in \ell$, by Pascal's theorem, it follows that $M = BV \cap DS \in \ell$. Furthermore, consider hexagon $ABVCDS$. We have that $AB \cap CD = K \in \ell$ and $BV \cap DS = M \in \ell$, so by Pascal's theorem once again, we have that $N = AS \cap CV \in \ell$. Finally, we turn to the hexagon $VUASRC$, where $AU \cap CR = L \in \ell$ and $AS \cap CV = N \in \ell$. Pascal's theorem, applied one last time, yields that UV and RS meet on ℓ. The conclusion follows.

Problem 11. (BMO 2008) Given a scalene acute triangle ABC with $AC > BC$ let F be the foot of the altitude from C. Let P be a point on AB, different from A so that $AF = PF$. Let H, O, M be the orthocenter, circumcenter and midpoint of side AC. Let X be the intersection point of BC and HP. Let Y be the intersection point of OM and FX and let OF intersect AC at Z. Prove that F, M, Y, Z are concyclic.

Solution. Denote by C' the second intersection of CH with the circumcircle of triangle ABC. Since F is the midpoint of the segment HC', the quadrilateral $AC'PH$ is a parallelogram, and F is its center of symmetry. Since X lies on line PH, its reflection X' into F lies on AC'. Thus, according to the Butterfly Theorem for the cyclic quadrilateral $AC'BC$ with circumcenter O, we have that $OF \perp XF$; combining this with $OM \perp AC$, we conclude that the points F, M, Y, Z are concyclic, as desired.

■

Comment. There are of course several more direct solutions using computations. For instance, take coordinates $F(0,0)$, $A(a,0)$, $B(0,-b)$, $C(0,c)$, $a, b, c, > 0$ and so $P(-a,0)$. Then $H(0,h)$, with $\frac{h}{a} = \frac{b}{c}$, hence $h = \frac{ab}{c}$. For the point $O(\frac{a-b}{2}, o)$, we have

$$\left(\frac{a+b}{2}\right)^2 + o^2 = \left(\frac{a-b}{2}\right)^2 + (o-c)^2,$$

so $o = \frac{c^2-ab}{2c}$, while the point X is the intersection of the lines

$$HP : -\frac{x}{a} + \frac{y}{h} = 1, \quad BC : -\frac{x}{b} + \frac{y}{c} = 1,$$

whence

$$X\left(\frac{b(ab-c^2)}{c^2-b^2}, \frac{bc(a-b)}{c^2-b^2}\right).$$

Thus,

$$\overrightarrow{FO} \cdot \overrightarrow{FX} = \frac{b(ab - c^2)}{c^2 - b^2} \cdot \frac{a - b}{2} + \frac{bc(a - b)}{c^2 - b^2} \cdot \frac{c^2 - ab}{2c} = 0,$$

which proves that $FO \perp FX$, as before.

Another possibility is to show that $\angle OFE = \angle CFX$ (E is the midpoint of side AB) by means of trigonometry. We have

$$\tan OFE = \frac{OE}{EF} = \frac{OA \cos C}{\frac{c}{2} - a \cos B} = \frac{\cos C}{\sin C - 2 \sin A \cos B} = \frac{\cos C}{\sin (B - A)},$$

since $\sin C - 2 \sin A \cos B = \sin (A + B) - 2 \sin A \cos B = \sin (B - A)$. Then, from Trigonometric Ceva in triangle PCF,

$$\frac{\sin CFX}{\sin BFX} \cdot \frac{\sin BPX}{\sin CPX} \cdot \frac{\sin PCX}{\sin FCX} = 1.$$

Now, $\angle BPX = \angle BAH = 90° - \angle B = \angle FCX$, $\angle CPX = \angle CAH = 90° - \angle C$, $\angle PCX = \angle PCF - \angle FCB = 90° - \angle A - (90° - \angle B) = \angle B - \angle A$. It follows that

$$\tan CFX = \frac{\sin CFX}{\sin BFX} = \frac{\sin CPX}{\sin PCX} = \frac{\sin (90° - C)}{\sin (B - A)} = \frac{\cos C}{\sin (B - A)} = \tan OFE,$$

where the latter equality holds from what we computed above. This proves that $\angle OFE = \angle CFX$, as desired.

■

Comment. We will generally avoid presenting computational solutions throughout this book. However, we do not want to ignore the fact that knowing analytical techniques can be incredibily useful in contests, if used properly. Consequently, we will make a compromise and include a few such solutions, with the intention of showing the reader that it is not always easy to reduce geometry problems to algebraic manipulations. The more difficult problems will always require a constant interplay between these computations and the visualization of the underlying geometric properties.

Problem 12. (Mehmet Sahin, Harvard College Mathematics Review) Let ABC be an arbitrary triangle and let I be the incenter of ABC. Let D, E, F be the points on lines BC, CA, AB respectively such that $\angle BID = \angle CIE = \angle AIF = 90°$, and define the following measurements: r_a, r_b, r_c are the exradii of the triangle ABC, $[DEF]$ is the area of DEF, and $[ABC]$ is the area of ABC. Prove that

$$\frac{[DEF]}{[ABC]} = \frac{4r(r_a + r_b + r_c)}{(a + b + c)^2}.$$

Solution. Let $a = BC$, $b = CA$, $c = AB$. Denote by r and s, respectively, the inradius and semiperimeter of triangle ABC. It is well known that $r = \frac{S}{s}$. We have

$$AF = AI \sec\frac{A}{2} = r\csc\frac{A}{2}\sec\frac{A}{2} = \frac{2r}{\sin A} = 2\left(\frac{S}{s}\right)\left(\frac{bc}{2[ABC]}\right) = \frac{bc}{s}$$

and similarly $CE = \frac{ab}{s}$.

Hence $AE = AC - EC = b - \frac{ab}{s} = \frac{b(s-a)}{s}$ and the area of triangle AFE is:

$$\frac{(AF)(AE)\sin A}{2} = \frac{1}{2}\left(\frac{bc}{s}\right)\left(\frac{b(s-a)}{s}\right)\left(\frac{2[ABC]}{bc}\right) = \frac{b(s-a)S}{s^2}.$$

Similarly, the area of triangle BDF and CED are $\frac{c(s-b)S}{s^2}$ and $\frac{a(s-c)S}{s^2}$. Thus

$$\begin{aligned} S' = S &= \frac{b(s-a)S + c(s-b)S + a(s-c)S}{s^2} \\ &= \frac{2(ab+bc+ca) - (a^2+b^2+c^2)}{(a+b+c)^2}S. \end{aligned}$$

It remains therefore to show that

$$4r(r_a + r_b + r_c) = 2(ab+bc+ca) - (a^2+b^2+c^2).$$

Using the well-known results:

$$r_a = \frac{[ABC]}{s-a}, r_b = \frac{[ABC]}{s-b}, r_c = \frac{[ABC]}{s-c}$$

and Heron's formula for S, we see that:

$$\begin{aligned} 4r(r_a + r_b + r_c) &= \frac{4[ABC]}{s}\left(\frac{[ABC]}{s-a} + \frac{[ABC]}{s-b} + \frac{[ABC]}{s-c}\right) \\ &= \frac{4[ABC]^2((s-b)(s-c) + (s-c)(s-a) + (s-a)(s-b))}{s(s-a)(s-b)(s-c)} \\ &= 4((s-b)(s-c) + (s-c)(s-a) + (s-a)(s-b)) \\ &= 2(ab+bc+ca) - (a^2+b^2+c^2), \end{aligned}$$

as desired.

■

Problem 13. (Omran Kouba, American Mathematical Monthly) Let a, b, and c be the lengths of the sides opposite vertices A, B, and C, respectively, in a nonobtuse triangle. Let h_a, h_b and h_c be the corresponding lengths of the altitudes. Show that

$$\left(\frac{h_a}{a}\right)^2+\left(\frac{h_b}{b}\right)^2+\left(\frac{h_c}{c}\right)^2\geq\frac{9}{4},$$

and determine the cases of equality.

Solution. By scale invariance it suffices to restrict our attention to triangles of unit area. For such triangles we have

$$H:=\left(\frac{h_a}{a}\right)^2+\left(\frac{h_b}{b}\right)^2+\left(\frac{h_c}{c}\right)^2=\frac{4}{a^4}+\frac{4}{b^4}+\frac{4}{c^4}.$$

Let C be the largest angle, and consider varying a and b with C fixed.

Since the area is $\frac{1}{2}ab\sin C$, this amounts to fixing aB. We first compare H to the value H_0 obtained for the triangle with side lengths $(\sqrt{ab},\sqrt{ab},c_0)$, where

$$c_0=ab+ab-2ab\cos C=2ab(1-\cos C).$$

We observe that

$$c^2-c_0^2=(a^2+b^2-2ab\cos C)-2ab(1-\cos C)=(a-b)^2$$

and

$$c^2+c_0^2=(a+b)^2-4ab\cos C.$$

Consequently, we compute

$$\begin{aligned}
H-H_0 &= \frac{4}{a^4}+\frac{4}{b^4}+\frac{4}{c^4}-\frac{4}{a^2b^2}-\frac{4}{a^2b^2}-\frac{4}{c_0^4}\\
&= \frac{4(a^2-b^2)^2}{a^4b^4}+\frac{4(c^4-c_0^4)}{c^4c_0^4}\\
&= 4(a-b)^2\left[\frac{(a+b)^2}{a^4b^4}-\frac{(a+b)^2}{c^4c_0^4}+\frac{4ab\cos C}{c^4c_0^4}\right].
\end{aligned}$$

Now $60^\circ\leq C\leq 90^\circ$. Since

$$c^4c_0^4\geq c_0^8=16a^4b^4(1-\cos C)^4\geq a^4b^4$$

for $60^\circ\leq C\leq 90^\circ$, we see that $H-H_0\geq 0$ with equality if and only if $a=b$. Thus it suffices to look at isosceles triangles.

In this case, unit area gives

$$a^2=b^2=2\csc C\text{ and }c^2=c_0^2=4(1-\cos C)\csc C$$

so we compute

$$\begin{aligned} H_0 - \frac{9}{4} &= 2\sin^2 C + \frac{\sin^2 C}{4(1-\cos C)^2} - \frac{9}{4} \\ &= 2(1-\cos^2 C) + \frac{1+\cos C}{4(1-\cos C)} - \frac{9}{4} \\ &= \frac{2(2\cos C - 1)^2 \cos C}{4(1-\cos C)} \\ &\geq 0, \end{aligned}$$

with equality if and only if $C = 60^\circ$ or $C = 90^\circ$. Thus, we get equality for an equilateral triangle or an isosceles right triangle.

■

Comment. Alternatively, one can note that

$$\frac{h_a}{a} = \frac{1}{\cot B + \cot C}$$

and substitute $x = \cot A$, $y = \cot B$, and $z = \cot C$, to reduce the problem to

$$\frac{1}{(x+y)^2} + \frac{1}{(y+z)^2} + \frac{1}{(z+x)^2} \geq \frac{9}{4},$$

where $xy + yz + zx = 1$, which is a famous inequality from the 1996 Iranian Mathematical Olympiad.

Problem 14. (Warut Suksompong and Potcharapol Suteparuk, IMO 2013) Let ABC be an acute-angled triangle with orthocenter H, and let W be a point on the side BC. Denote by M and N the feet of the altitudes from B and C, respectively. Denote by ω_1 the circumcircle of BWN, and let X be the point on ω_1 which is diametrically opposite to W. Analogously, denote by ω_2 the circumcircle of triangle CWM, and let Y be the point on ω_2 which is diametrically opposite to W. Prove that X, Y and H are collinear.

Solution 1. Let L be the foot of the altitude from A, and let Z be the second intersection point of circles ω_1 and ω_2, other than W. We show that X, Y, Z and H lie on the same line. Due to $\angle BNC = \angle BMC = 90^\circ$, the points B, C, N and M are concyclic; denote their circle by ω_3. Observe that the line WZ is the radical axis of ω_1 and ω_2; similarly, BN is the radical axis of ω_1 and ω_3, and CM is the radical axis of ω_2 and ω_3. Hence $A = BN \cap CM$ is the radical center of the three circles, and therefore WZ passes through A.

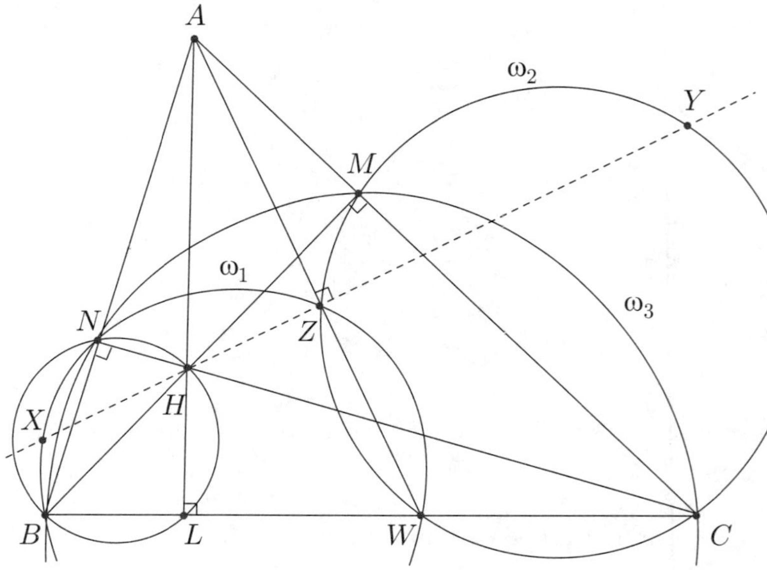

Since WZ and WY are diameters in ω_1 and ω_2, respectively, we have $\angle WZX = \angle WZY = 90°$, so the points X and Y lie on the line through Z, perpendicular to WZ.

The quadrilateral $BLHN$ is cyclic, because it has two opposite right angles. From the power of A with respect to the circles ω_1 and $BLHN$ we find $AL \cdot AH = AB \cdot AN = AW \cdot AZ$. If H lies on the line AW, then this implies $H = Z$ immediately. Otherwise, by $\frac{AZ}{AH} = \frac{AL}{AW}$ the triangles AHZ and AWL are similar. Then $\angle HZA = \angle WLA = 90°$, so the point H also lies on the line XYZ.

The above approach was perhaps the approach that most popular among contestants. In what follows, however, we give ten other different proofs to illustrate a few ideas and theorems that will resurface throughout the book. The first two are simplifications of the argument given above:

Solution 2. Let Z be the second intersection point of ω_1 and ω_2. Note that the points A, Z and W are collinear, as in the first solution. Furthermore, also note that the point Z lies on the line XY (as in the first solution). The next step is to see that Z lies on the circumcircle of quadrilateral $ANHM$, which has diameter AH. This is however immediate. Note that Miquel's theorem, applied for the points N, W, M, on the sides AB, BC, CA, respectively, yields that circles ω_1, ω_2 and (ANM) are concurrent. Hence, $ANZM$ is cyclic. Let Ω be its circumcircle. Since $\angle HNA = \angle HMA = 90°$, H lies on Ω as well, and AH is a diameter of Ω.

Since XW is the diameter of ω_1, we have that $\angle XZW = 90°$. Since AH is the diameter of ω_3, we have that $\angle HZA = 90°$. It follows that $\angle HZW = 90°$, so X, H and Z lie on a line.

Solution 3. One can easily prove that M, N, B, C are concyclic and that A, M, H, N are concyclic. So, by Power of Point, $AN \cdot AB = AM \cdot AC$. Now, consider the inversion with center A and power $AN \cdot AB$ so that ω_1 and ω_2 are fixed. As usual, we denote by X' the image of X under this inversion. One can see that $M' = C$, $C' = M$, $N' = B$, $B' = N$ and circle $AMHN$ maps to line BC.

Point W lies on both ω_1 and ω_2, which remain fixed, so W' is precisely the second intersection point of ω_1 and ω_2 (point Z in the first solution). Finally, since W lies on the line BC, Z lies on circle $(AMHN)$, which has diameter AH, so $AZ \perp ZH$ and we finish like in the first solution.

Solution 4. Let $D = BX \cap CN$. We have that $\angle DNB$ and $\angle BNC$ are right angles and $\angle BDN = \angle NBC$, so triangles BND and CNB are similar.

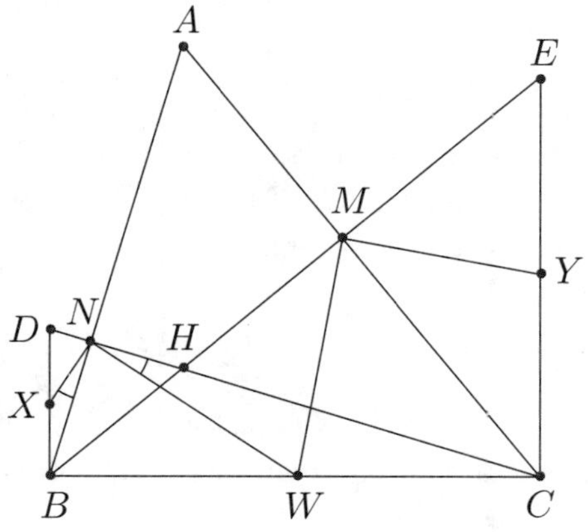

Since $\angle XNW = \angle BNC = 90°$, we have that $\angle XNB = \angle WNC$, hence the aforementioned similarity maps X to W. Therefore

$$\frac{BX}{XD} = \frac{CW}{WB}.$$

Similarly, if we define $E = CY \cap BH$,

$$\frac{EY}{YC} = \frac{CW}{WB}.$$

Hence $\frac{BX}{XD} = \frac{EY}{YC}$. Since BD and CE are parallel (both perpendicular to BC), it follows from the similarity of the triangles BDH and ECH, which maps X to Y, that X, H and Y are collinear.

Solution 5. Let B' and C' be the intersections of the line through W orthogonal to BC and the lines CN and BM, respectively. The quadrilaterals

$WB'NB$ and $WC'MC$ are both cyclic since they have two opposite right angles. By the definition of X and Y and the constructions of B' and C', the quadrilaterals $WB'XB$ and $WC'YC$ are rectangles.

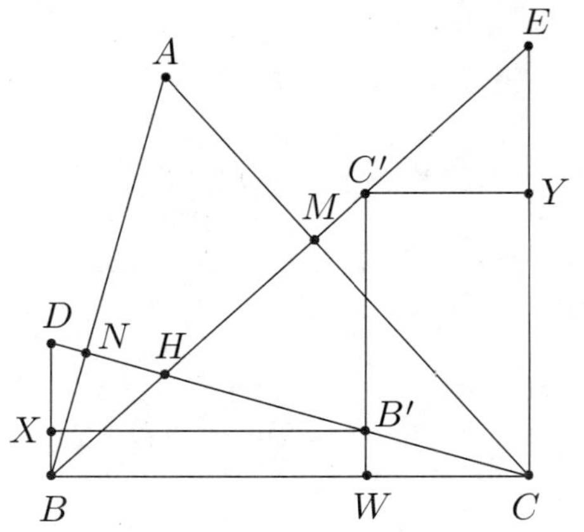

Let D be the intersection of the lines BX and CN, and let E be the intersection of the lines CY and BM. Since the lines BX, WB' and CY are all parallel, the following triangles are similar

$$\triangle HBD \sim \triangle HEC, \quad \triangle CB'W \sim \triangle CDB, \quad \triangle C'EY \sim \triangle BEC.$$

Now look at the homothety with center H that takes the triangle HBD to the triangle HEC. Since

$$\frac{BX}{BD} = \frac{B'W}{BD} = \frac{CW}{BC} = \frac{YC'}{BC} = \frac{YE}{CE},$$

this homothety takes X to Y. Hence H, X and Y are collinear.

Solution 6. Since the quadrilateral $BXNW$ is cyclic, $\angle WXN = \angle B$. Then $\triangle XNW \sim \triangle HNA$, so

$$\frac{XN}{NW} = \frac{HN}{AN}.$$

Note also that $\angle HNX = \angle ANW$. It follows that $\triangle HNX \sim \triangle ANW$. Thus $\angle NHX = \angle NAW$. Analogously, $\angle MHY = \angle MAW$. Moreover, $\angle NAW + \angle MAW = \angle A$. Then $\angle NHX + \angle MHY = \angle A$.

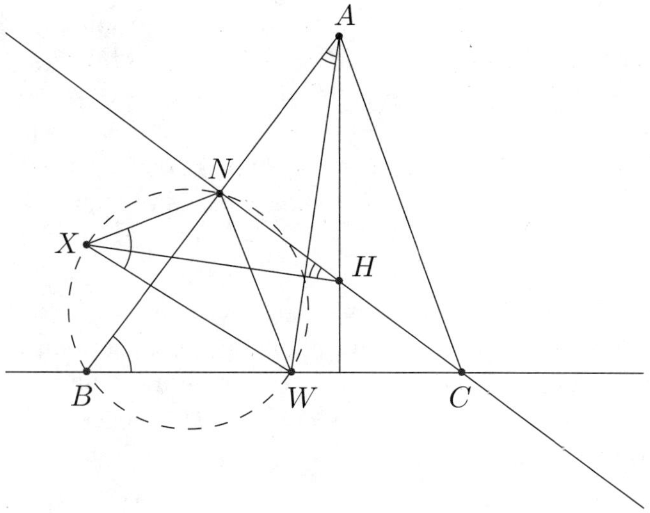

Finally, since the quadrilateral $ANHM$ is cyclic, we have that $\angle NHM = 180^\circ - \angle A$. It follows that $\angle NHX + \angle NHM + \angle MHY = 180^\circ$, so X, H, Y are collinear.

Solution 7. The following solution is actually very similar to the previous one; just scale it by $\frac{1}{2}$ around W.

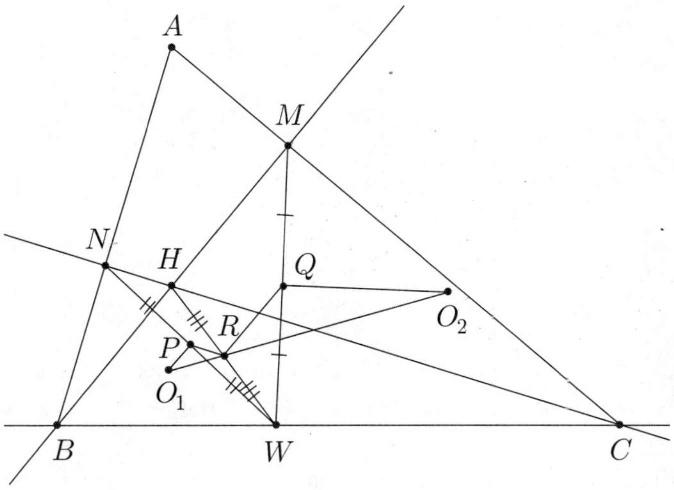

Let O_1 and O_2 be the centers of the circles ω_1 and ω_2 respectively. Let P, Q and R be the midpoints of the segments WN, WH and WM respectively. It is sufficient to show that O_1, R and O_2 are collinear.

Note that PR is parallel to NH, so $\angle WPR = \angle WNH$, and so $\angle O_1PR = \angle WNA$. Also

$$\frac{O_1P}{NW} = \frac{1}{2}\frac{O_1P}{PW} = \frac{1}{2}\frac{NH}{AN} = \frac{PR}{AN}.$$

Thus $\triangle O_1PR \sim \triangle WNA$, so $\angle PRO_1 = \angle WAN$. Analogously, $\angle WAM = \angle QRO_2$. Moreover, $\angle PRQ = \angle NHM = 180° - \angle A$, thus $O_1RO_2 = 180°$, so the result follows.

The next three solutions appeal to theorems that seem completely unrelated with the configuration we are dealing with, and they are quite remarkable.

Solution 8. Consider the circle ω_3 with diameter BC, which passes through M and N. Let O be the center of this circle, which happens to be also the midpoint of BC. Let O_1 and O_2 be the centers of ω_1 and ω_2, respectively. BN is a common chord of ω_3 and ω_1, so $OO_1 \perp BN$; since $CN \perp AB$ as well, $CN \parallel OO_1$. Analogously, $BM \parallel OO_2$.

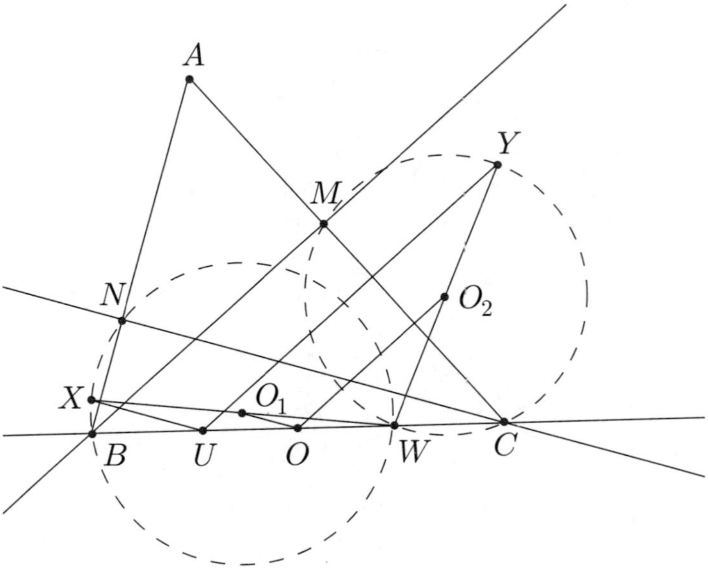

Now let U be the reflection of W through O. Since O_1 is the midpoint of WX and O is the midpoint of WU, $UX \parallel OO_1$, which means that $UX \parallel CN$. Analogously, $UY \parallel BM$.

We finish the problem invoking Pappus' theorem on line BCU and the line at infinity, which contains I_C, the point where all lines perpendicular to BN meet, I_B, the point where all lines perpendicular to CM meet, and I_U, the point where all lines perpendicular to BC, including BX and CY, meet. So $BI_B \cap CI_C = H$, $BI_U \cap UI_B = X$ and $CI_U \cap UI_C = Y$ are collinear.

Solution 9. As in the first solution, we begin by noticing that A lies on the line WZ as in the first solution, and continue from there.

Let B' and C' be the second intersections of BM with ω_2 and CN with ω_1, respectively. Note that, since $\angle BNC' = 90°$ then $\angle BWC' = 90°$ and thus BX is parallel to WC'. Call I their intersection at infinity. Let $H' = XZ \cap C'N$.

By applying Pascal's theom on the cyclic hexagon $BXZWC'N$, the points I, H' and A are collinear. That is, $H'A$ is also perpendicular to BC. Hence $H' = H$.

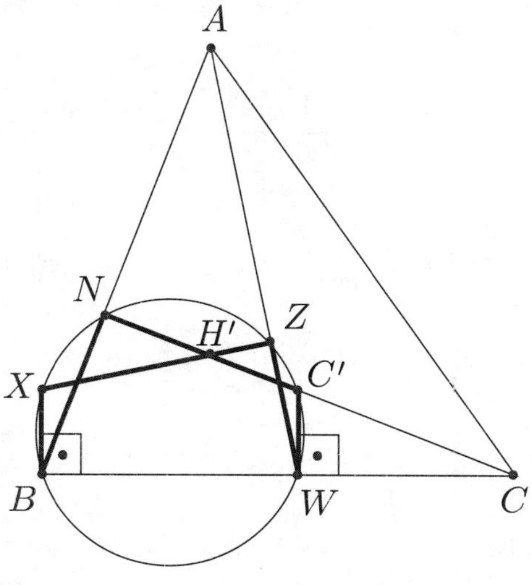

This implies that X, H and Z are collinear. Similarly, we prove that Y, H and Z are collinear. Hence X, Y and H are collinear.

Solution 10. Let, as in the previous solution, B' and C' be the second intersections of BM with ω_2 and CN with ω_1, respectively, and note that $B' \in BM$ and $C' \in CN$. By the Newton-Gauss theorem (which we review in the solution of Problem 71), applied for quadrilateral $BWC'H$, the midpoints X', H', Y' of segments BC', HW, and $B'C$, respectively, are collinear. However, BC' and $B'C$ are diameters in ω_1 and ω_2, thus X' and Y' are in fact the centers of ω_1 and ω_2, respectively. Hence, the homothety with center W and ratio 2 maps X' to X, Y' to Y and H' to H, allowing us to conclude that points X, Y, H are collinear, as claimed.

Last but not least, we would like to include a computational approach that is instructive.

Solution 11. Let $a = BC$, $b = AC$, $c = AB$ be the side lengths of triangle ABC and let $\alpha = \angle BAC$, $\beta = \angle ABC$ $\gamma = \angle ACB$. We have $\angle WBX =$

$\angle YCW = 90°$. By the law of cosines on triangle BWN, since $BN = a\cos\beta$,

$$\begin{aligned} WN^2 &= BN^2 + BW^2 - 2BN\cdot BW\cdot\cos\beta \\ &= a^2\cos^2\beta + BW^2 - 2a\cdot BW\cdot\cos^2\beta \\ &= (a-BW)^2\cos^2\beta + BW^2\sin^2\beta \\ &= CW^2\cos^2\beta + BW^2\sin^2\beta. \end{aligned}$$

Applying the Law of Sines in the same triangle, whose circumcircle has diameter XW,

$$\begin{aligned} XW^2 &= \frac{WN^2}{\sin^2\beta} \\ &= CW^2\cot^2\beta + BW^2, \end{aligned}$$

hence $BX^2 = CW^2\cot^2\beta$, i.e. $BX = CW\cot\beta$. Analogously, we get that $CY = BW\cot\gamma$.

Let the foot of the altitude of A on BC be L. By comparing slopes, the problem is equivalent to

$$\frac{CY - HL}{CL} = \frac{HL - BX}{BL}.$$

Comparing slopes is always very useful when proving collinearities analytically. Here, triangles CHL and CBN are similar, so

$$HL = \frac{BN\cdot CL}{CN} = \frac{a\cos\beta\cdot b\cos\gamma}{a\sin\beta} = b\cos\gamma\cot\beta.$$

Analogously, $HL = c\cos\beta\cot\gamma$. So, since $CL = b\cos\gamma$ and $BL = c\cos\beta$,

$$\frac{CY - HL}{CL} = \frac{BW\cot\gamma - c\cos\beta\cot\gamma}{b\cos\gamma} = \frac{BW - BL}{b\sin\gamma} = \frac{WL}{b\sin\gamma}$$

and, similarly,

$$\frac{HL - BX}{BL} = \frac{WL}{c\sin\beta}.$$

This completes the proof, since the Law of Sines in triangle ABC assures that $\frac{b}{\sin\beta} = \frac{c}{\sin\gamma}$.

■

Problem 15. (Titu Andreescu, USAMO 2014 Shortlist) Let ABC be a triangle and let P be a point in its interior. Let X, Y, Z be the intersections of AP, BP, CP with sides BC, CA, AB, respectively. Prove that

$$\frac{XB}{XY}\cdot\frac{YC}{YZ}\cdot\frac{ZA}{ZX} \le \frac{R}{2r}.$$

Solution. By the Law of Cosines and the AM-GM inequality,

$$\begin{aligned} YZ^2 &= AY^2 + AZ^2 - 2AY \cdot AZ \cdot \cos A \\ &\geq 2AY \cdot AZ(1 - \cos A) \\ &= 4AY \cdot AZ \sin^2 \frac{A}{2}. \end{aligned}$$

Similarly, $ZX^2 \geq 4BZ \cdot BX \sin^2 \frac{B}{2}$ and $XY^2 \geq 4CX \cdot CY \sin^2 \frac{C}{2}$. Consequently,

$$\begin{aligned} (XY \cdot YZ \cdot ZX)^2 &\geq 64 \prod_{A,B,C} (AY \cdot AZ) \cdot \prod_{A,B,C} \sin^2 \frac{A}{2} \\ &= \frac{4r^2}{R^2} \cdot \prod_{A,B,C} (AY \cdot AZ), \end{aligned}$$

where the latter equality holds because of the well-known identity:

$$\sin \frac{A}{2} \sin \frac{B}{2} \sin \frac{C}{2} = \frac{r}{4R}.$$

On the other hand, lines AX, BY, CZ are concurrent at P, so Ceva's theorem yields

$$\frac{XB}{XC} \cdot \frac{YC}{YA} \cdot \frac{ZA}{ZB} = 1.$$

Hence, we conclude that

$$\begin{aligned} \left(\frac{XY}{XB} \cdot \frac{YZ}{YC} \cdot \frac{ZX}{ZA} \right)^2 &= \frac{(XY \cdot YZ \cdot ZX)^2}{AY \cdot AZ \cdot BZ \cdot BX \cdot CX \cdot CY} \\ &\geq \frac{4r^2}{R^2}, \end{aligned}$$

by the above. It follows that

$$\frac{XY}{XB} \cdot \frac{YZ}{YC} \cdot \frac{ZX}{ZA} \geq \frac{2r}{R},$$

as desired.

■

Problem 16. (Vladimir Protassov) Let $\mathcal{C}_1$ be a circle that is tangent to sides AB and AC of triangle ABC and let $\mathcal{C}_2$ be a circle through B and C that is tangent to $\mathcal{C}_1$ at point D. Prove that the incenter of triangle ABC lies on the internal angle bisector of $\angle BDC$.

Solution. Consider the line through the center of $\mathcal{C}_1$ which intersects $\mathcal{C}_1$ again at X, closer to A. Furthermore, consider the line through the center of $\mathcal{C}_2$ which intersects $\mathcal{C}_2$ again at Y, further from A. Clearly, X, D, and Y are collinear. Let E and F be the tangency points of $\mathcal{C}_1$ with AC and AB, respectively. Extend XF to meet YB at M, and extend XE to meet YC at N. From $\angle XFD = \angle DCY = \angle MBD$, we get that $BMFD$ is cyclic, so let us proceed by taking I' to be the second intersection this circle with XY. We have that $\angle I'BA = \angle FDX = \frac{1}{2}\angle ABC$.

Similarly, let the circumcircle of $DENC$ intersect XY again at I''; then $\angle I''CA = \frac{1}{2}\angle ACB$. To prove that $I' \equiv I''$, we proceed as follows. From $\angle BI'Y = \angle BFD = \angle FXD$, we get $MX \parallel BI'$, and from $\angle BMI' = \angle BDI' = \angle YBC$, we get $BC \parallel MI'$. So, if we let XY intersect BC at P, we have

$$\left(\frac{YP}{YI'}\right)^2 = \frac{YP}{YI'}\frac{YB}{YM} = \frac{YP}{YI'}\frac{YI'}{YD} = \frac{YP}{YD}.$$

Analogously,

$$\left(\frac{YP}{YI''}\right)^2 = \frac{YP}{YD}.$$

Thus, it follows that $I' \equiv I''$, as desired.

■

Comment. If $\mathcal{C}_2$ is the circumcircle of triangle ABC, circle $\mathcal{C}_1$ is called the *A-mixtilinear incircle* of ABC. It's a circle that has numerous beautiful properties involving the incircle. We will see a couple of them later.

Problem 17. (Alexey Zaslavsky, Hyacinthos Newsgroup) Let ABC be a triangle and let P, Q be two points lying on its circumcircle. Prove that their Simson lines meet on the A-altitude of triangle ABC if and only if $PQ \| BC$.

Solution. We begin by referring to a classical result about the Simson line that we shall use here.

Lemma. The Simson line of a point P lying on the circumcircle of triangle ABC passes through the midpoint of the segment PH, where H denotes the orthocenter of ABC.

There are quite a few ways to think about this result, ranging from synthetic to completely computational. We give the short synthetic one below.

Proof. The idea is to take the reflection A' of the orthocenter H across BC. We know that it must lie on the circumcircle. Furthermore, let PA' meet BC at E

and let D be the foot of the altitude from A and X, Y be the projections of P onto the sidelines BC and CA. Since triangle HEA' is isosceles, we have that $\angle HEB = \angle A'EB$. But on the other hand, we have that $\angle YXB = \angle YPC$, as $YXCP$ is cyclic; hence $\angle YXB = 90° - \angle PCA = 90° - \angle A'AP = \angle A'EB$ (as $A'CPA$ is cyclic). Thus, we conclude that $\angle HEB = \angle YXB$, so the Simson line XY is parallel to HE. However, triangle PEX has a right angle at X and has the median XM so that $XM = ME$, as triangle MEX is isosceles because of the angle equality at the base (here M denotes the midpoint of PE). Thus, the Simson line XY coincides with the P-midline of triangle PEH, and so it bisects the segment PH, as desired. This proves the Lemma.

□

Returning to the problem, let X, Y be the projections of P and Q on the line BC and let U, V be the midpoints of segments HP, HQ, respectively, where H denotes the orthocenter of ABC. Recall that the Simson line of a point bisects the segment joining that point with the orthocenter H; thus, we know that the Simson lines of P and Q are precisely the lines XU and YV. Hence, with this observation in mind, we see that Desargues' theorem for triangles XPU and YQV yields that lines PQ, UV, BC are concurrent if and only if the intersections $XP \cap YQ = \infty$, $PU \cap QV = H$, $XU = s_P \cap s_Q = YV$ are collinear. But the line UV is the H-midline in triangle HPQ, so $UV \| PQ$; thus we immediately get the conclusion that PQ, UV, BC are concurrent if and only if $PQ \| BC$, which from Desargues' theorem happens if and only if the Simson lines intersect on the line determined by H and the "intersection point" of the lines XP and YQ which are both perpendicular to BC. This completes the proof, as the latter just means that s_P and s_Q have to intersect on the A-altitude.

■

Problem 18. (MOSP) Let ABC be a triangle and let P be a point in its interior with pedal triangle DEF. Suppose that that the lines DE and DF are perpendicular. Prove that the isogonal conjugate of P with respect to ABC is the orthocenter of triangle AEF.

Solution. Recall the following basic properties of isogonal conjugates.

Lemma 1. Let P be a point in the plane of the triangle ABC and let X, Y, Z be the projections of P onto the sidelines BC, CA, and AB, respectively. Then, the perpendiculars from the vertices A, B, C to the lines YZ, ZX, XY are concurrent at the isogonal conjugate of P with respect to ABC.

Lemma 2. Let P be a point in the plane of the triangle ABC and let X', Y', Z' be the reflections of the point P in the sidelines BC, CA, and AB, respectively. Then, the isogonal conjugate of P with respect to ABC is the circumcenter of triangle $X'Y'Z'$.

The first one is a one-line angle chase, which we leave for the reader. We prove the second one.

Proof of Lemma 2. Let X, Y, Z be the midpoints of segments PX, PY, PZ. Obviously, triangle XYZ is the pedal triangle of P with respect to ABC and is homothetic with triangle $X'Y'Z'$.

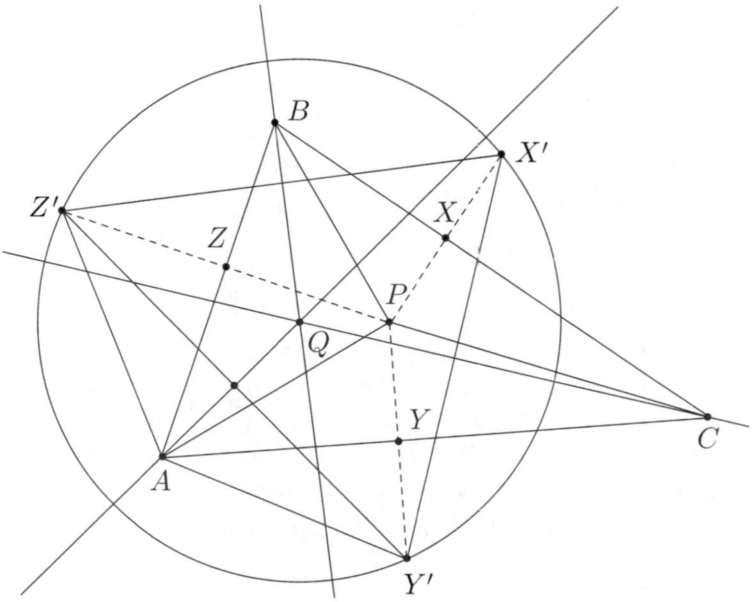

By Lemma 1, we immediately get that if Q denotes the isogonal conjugate of P, then $AQ \perp Y'Z'$. However, we still need to show that this is the perpendicular bisector of $Y'Z'$ or in other words that A lies on the perpendicular bisector. But this is clear! Note that Y' is the reflection of P across AC, so $AP = AY'$, and Z' is the reflection of P across AB, so $AP = AZ'$; hence $AY' = AZ'$, i.e., triangle $AY'Z'$ is isosceles and so A needs to lie on the perpendicular bisector of $Y'Z'$ as claimed. This completes the proof, since doing the same thing for BQ and CQ, we conclude that they are the perpendicular bisectors of the sides of triangle $X'Y'Z'$, so they need to meet at the circumcenter of $X'Y'Z'$.

$\square$

Returning to the problem, let Q be the isogonal conjugate of P with respect to ABC. Let X, Y, Z be the reflections of P across the sidelines BC, CA, and AB, respectively. By Lemma 2, Q is the circumcenter of triangle XYZ.

Now, triangles DEF and XYZ are clearly similar; hence, since DE and DF are perpendicular to each other, so are XY and XZ. It follows that Q is the midpoint of YZ.

Since M, E, F are the midpoints of the sides of triangle PZY, we have $EM \| PZ$ and $FM \| PY$. But $PZ \perp AF$ and $PY \perp AE$. We conclude that $EM \perp AF$ and $FM \perp AE$, so M is the orthocenter of triangle AEF. This completes the proof.

■

Problem 19. (Luis Gonzalez, AOPS Forum) Let τ be a line that is tangent to the circumcircle $\Gamma(O, R)$ of triangle ABC. Let I, I_a, I_b, I_c be the incenter and the three excenters of ABC and let $\delta(P)$ stands for the distance from point P to τ. Prove that there exists a choice of signs so that the following identity is true:

$$\pm\delta(I) \pm \delta(I_a) \pm \delta(I_b) \pm \delta(I_c) = 4R.$$

Solution. We begin with a simple fact about the orthocenter and the nine-point center.

Lemma. For any triangle ABC with orthocenter H and nine-point center N,

$$\overrightarrow{NH} + \overrightarrow{NA} + \overrightarrow{NB} + \overrightarrow{NC} = 0.$$

Proof of Lemma. Let G be the centroid then $\frac{GN}{NH} = \frac{1}{3}$.
Since $\overrightarrow{GA} + \overrightarrow{GB} + \overrightarrow{GC} = 0$, we have

$$\begin{aligned}\overrightarrow{NH} + \overrightarrow{NA} + \overrightarrow{NB} + \overrightarrow{NC} &= \overrightarrow{GN} + \overrightarrow{NA} + \overrightarrow{GN} + \overrightarrow{NB} + \overrightarrow{GN} + \overrightarrow{NC}\\ &= \overrightarrow{GA} + \overrightarrow{GB} + \overrightarrow{GC} = 0.\end{aligned}$$

This proves the Lemma.

□

Going back to the question, we note that O is the nine point center of triangle $I_aI_bI_c$ and I is the orthocenter. Then, by the above, we have that $\overrightarrow{OI} + \overrightarrow{OI_a} + \overrightarrow{OI_b} + \overrightarrow{OI_c} = 0$. So consider the line ℓ that passes through O and is parallel to τ. Let $f(P)$ be the vector from P to ℓ. Then, $\pm f(I) \pm f(I_a) \pm f(I_b) \pm f(I_c) = 0$. We define $\delta(P)$ to be the vector from P to τ, which gives $\delta(I) = f(I) \pm R$; consequently, we have

$$|\delta(I) + \delta(I_a) + \delta(I_b) + \delta(I_c)| = \pm 4R.$$

Of course, δ is a vector, so we can assign signs to get the desired result.

■

Problem 20. (Feuerbach's Theorem) Prove that the nine-point circle of a triangle is tangent to the incircle and the excircles of the triangle.

Solution 1. We will give two different solutions that are very instructive. The first one is by inversion and comes in a few steps.

Claim 1. Let ABC be a triangle and let Ψ be the inversion with center A and power k^2. If $\mathcal{C}(O,R)$ denotes the circumcircle of ABC, then $\Psi(\mathcal{C}(O,R))$ is a line L which is antiparallel to the line BC.

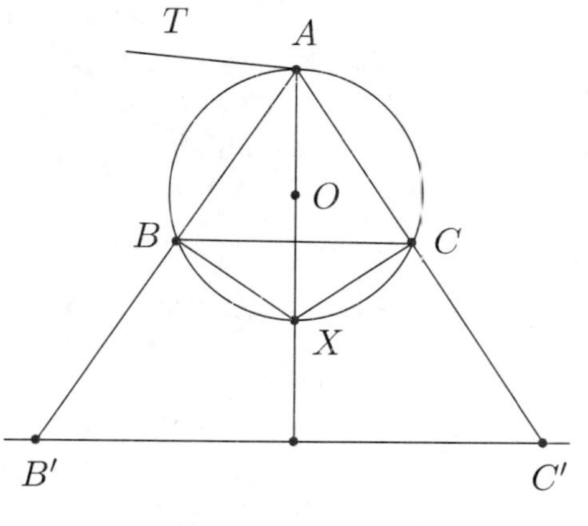

Proof. This follows immediately from the basic properties of inversion.

We know that $\Psi(\mathcal{C}(O,R)) = B_1C_1$, where B_1, C_1 are the images of B and C under Ψ. We have that

$$AB \cdot AB_1 = AC \cdot AC_1 = k^2,$$

so B_1C_1CB is cyclic, i.e., $\angle ABC = \angle AC_1B_1$, and so the lines B_1C_1 and BC are antiparallel.

Claim 2. In a triangle ABC, let A_1, A_2, A_3 and P be the following points on the side BC: A_1 is the midpoint, A_2 is the foot of the altitude from A, A_3 is the point where the internal angle bisector of angle A meets BC, P is the foot of the perpendicular from I to BC (i.e., the tangency point of BC with the incircle). Then, $A_1P^2 = A_1A_2 \cdot A_1A_3$.

Proof. Without loss of generality, assume that $a \geq b \geq c$. Indeed, we have $A_1P = A_1B - BP = \frac{a}{2} - (s-b) = \frac{b-c}{2}$, where s is he semiperimeter of ABC. Also, $A_1A_2 = \frac{b^2-c^2}{2a}$, and

$$A_1A_3 = BA_1 - BA_3 = \frac{a}{2} - \frac{ac}{b+c} = \frac{a(b-c)}{2(b+c)}.$$

Thus, it follows that

$$A_1P^2 = A_1A_3 \cdot A_1A_2 = \left(\frac{b-c}{2}\right)^2,$$

as claimed.

[Alternatively, one can be sophisticated about this as follows: Let A_4 be the tangency point of the A-excircle with side BC, and note that the quadruplet (A_2, P, A_3, A_4) a harmonic division. Indeed, this follows from the classical

Nagel's Lemma. Let ABC be a triangle and let P and Q be the points where the incircle and A-excircle touch BC, respectively. Let P' be the antipode of point P with respect to the incircle. Then, points A, P', Q are collinear.

As A_1 is the midpoint of segment PA_3, the identity now follows from the Lemma about harmonic divisions from the solution of Problem 71.]

Returning to the proof of Feuerbach's theorem, let A_1 be the midpoint of BC and let P and P_a be the points of tangency of the incircle and the A-excircle with the side BC. Consider the inversion Ψ with center A_1 and power $k^2 = A_1P^2$.

Since P is the point of tangency of the side BC with the incircle $\mathcal{C}(I,r)$, we have that A_1P^2 is the power of A_1 with respect to $\mathcal{C}(I,r)$, so $\Psi(\mathcal{C}(I,r)) = \Psi(I,r)$! Also, since P_a is the point of tangency of the side BC with the A-excircle $\mathcal{C}(I_a,r_a)$ and $A_1P_a = A_1P$, then $A_1P_a^2 = A_1P^2 = k^2$ is also the power of A_1 with respect to $\mathcal{C}(I_a,r_a)$, so $\mathcal{C}(I_a,r_a)$ is also fixed under Ψ. That is, both $\mathcal{C}(I,r)$ and $\mathcal{C}(I_a,r_a)$ are invariant under Ψ.

Now, consider the image of the nine-point circle under Ψ. First of all, since A_1 belongs to the nine-point circle, which we denote by $\mathcal{C}_9$, we first infer that $\Psi(\mathcal{C}_9)$ is a line d. But $\mathcal{C}_9$ is the circumcircle of the triangle with vertices at the midpoints A_1, B_1, C_1 of the sides of ABC, so the line d is antiparallel to the line B_1C_1 by Claim 1. However, B_1C_1 is parallel to BC, so d is actually antiparallel to BC. Furthermore, we also have that $A_1A_2 \cdot A_1A_3 = A_1P^2$ by Claim 2, and so, since A_2 belongs to $\mathcal{C}_9$, we get that d is a line which passes through A_3, as $\Psi(A_2) = A_3$.

Next, let $B'C'$ be the second common tangent of the two circles $\mathcal{C}(I,r)$ and $\mathcal{C}(I_a,r_a)$. Since A_3 is the foot of the bisector of the angle at A, these common tangents intersect at A_3. Now, we claim that $\angle ABC = \angle AC'B'$ and $\angle AB'C' = \angle ACB$. From this, it follows that the second common tangent $B'C'$ is antiparallel to BC. Since A_3 is on $B'C'$ then it follows that the line d must be $B'C'$, that is $\Psi(\mathcal{C}_9) = d$. Finally, since d is tangent to $\mathcal{C}(I,r)$ and $\mathcal{C}(I_a,r_a)$ then $\mathcal{C}_9 = \Psi^{-1}(d)$ is tangent to $\mathcal{C}(I,r)$ and $\mathcal{C}(I_a,r_a)$. Thus, $\mathcal{C}_9$ is tangent to the incircle and A-excircle. Similarly, it can be shown that $\mathcal{C}_9$ is also tangent to the other two excircles.

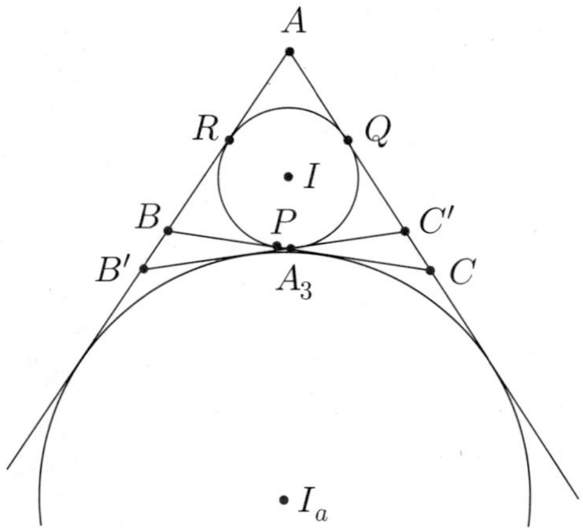

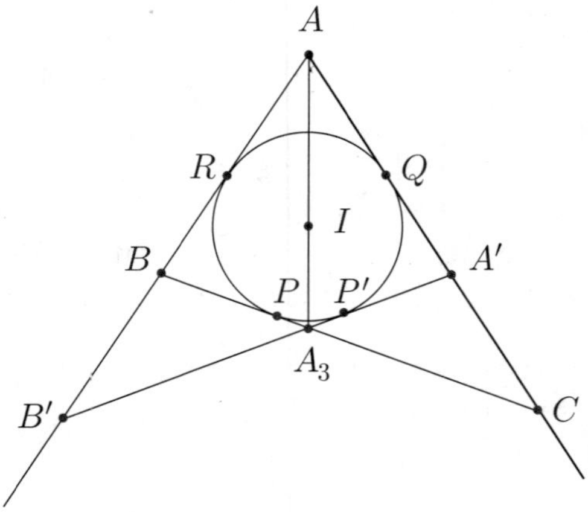

Hence, it remains to show that the common tangents BC and $B'C'$ are antiparallel. Let P, Q and R be the points of tangency of the sides BC, CA, and AB, with the incircle of the triangle, respectively. Let P' be the point of tangency of the second common tangent $B'C'$ with the incircle $\mathcal{C}(I,r)$. The triangles AIR and AIQ are similar so $\angle AIR = \angle AIQ$. Also, the triangles A_3IP and A_3IP' are similar, so $\angle A_3IP = \angle A_3IP'$. Then,

$$\angle PIR = 180^\circ - (\angle AIR + \angle A_3IP) = 180^\circ - (\angle AIQ + \angle A_3IP') = \angle P'IQ.$$

Since the quadrilaterals $PIRB$ and $P'IQC'$ are cyclic, we get that

$$\angle PBR = 180^\circ - \angle PIR = 180^\circ - \angle P'IQ = \angle P'C'Q,$$

i.e., $\angle ABC = \angle AC'B'$. Thus

$$\angle BCA = 180^\circ - (\angle A + \angle ABC) = 180^{circ} - (\angle A + \angle AC'B') = \angle AB'C'.$$

It follows that BC and $B'C'$ are antiparallel, as required, thus completing the proof.

Solution 2. The second proof uses the powerful

Casey's Theorem. Given four circles $\mathcal{C}_i$, $i = 1,2,3,4$, let t_{ij} be the length of a common tangent between $\mathcal{C}_i$ and $\mathcal{C}_j$. The four circles are tangent to a fifth circle (or line) if and only if, for an appropriate choice of signs, we have that

$$t_{12}t_{34} \pm t_{13}t_{42} \pm t_{14}t_{23} = 0.$$

The direct implication of the theorem follows immediately from Casey's Chord theorem, which we saw in the proof of Problem 80, combined with

Ptolemy's theorem. However, the converse is very intricate and all known proofs are very long and computational, so we shall omit it here. We refer to [1] for details.

To deduce Feuerbach's theorem now, we proceed as follows. First, let us change the notation a bit to emphasize how different these proofs are. Let the sides BC, CA, AB of triangle ABC have midpoints D, E, F respectively, and let Γ be the incircle of the triangle. Let a, b, c be the side lengths of ABC, and let s be its semiperimeter. We now consider the 4-tuple of circles $(D,\ E,\ F,\ \Gamma)$. Here is what we find:

$$t_{DE} = \frac{c}{2},\quad t_{DF} = \frac{b}{2},\quad t_{EF} = \frac{a}{2},$$

$$t_{D\Gamma} = \left|\frac{a}{2} - (s-b)\right| = \left|\frac{b-c}{2}\right|,$$

$$t_{E\Gamma} = \left|\frac{b}{2} - (s-c)\right| = \left|\frac{a-c}{2}\right|,$$

$$t_{F\Gamma} = \left|\frac{c}{2} - (s-a)\right| = \left|\frac{b-a}{2}\right|.$$

We need to check whether, for some combination of $+$, $-$ signs, we have

$$\pm c(b-a) \pm a(b-c) \pm b(a-c) = 0.$$

But this is immediate! According to Casey's theorem there exists a circle what touches each of D, E, F and Γ. Since the circle passing through D, E, F is the nine-point circle of the triangle, it follows that Γ and the nine-point circle are tangent to each other.

Analogously, we can use Casey's theorem to prove that the incircle Γ is tangent to each of the excircles of ABC.

■

Comment. The tangency point between the incircle and the nine-point circle is known as the Feuerbach point of the triangle ABC.

Problem 21. (Zuming Feng, USA IMO TST 2012) In acute triangle ABC, $\angle A < \angle B$ and $\angle A < \angle C$. Let P be a variable point on side BC. Points D and E lie on sides AB and AC, respectively, such that $BP = PD$ and $CP = PE$. Prove that as P moves along side BC, the circumcircle of triangle ADE passes through a fixed point other than A.

Solution 1. We will prove that the fixed point is the orthocenter H of triangle ABC. Let X be the foot of the perpendicular from C to AB, and let Y be the foot of the perpendicular from B to AC. Note that if M is the midpoint of BC, then $MB = MX = MY = MC$. Suppose without loss of generality that P is between B and M. Then D is between B and X, and E is between A and Y. The quadrilateral $AXHY$ is cyclic, since $\angle AXH = \angle AYH = 90°$. To show that $ADHE$ is also cyclic, it suffices to show that triangles DHX and EHY are similar, or that

$$\frac{DX}{EY} = \frac{XH}{YH}.$$

Applying the Law of Sines in the cyclic quadrilateral $AXHY$, we find that $\frac{XH}{YH} = \frac{\cos B}{\cos C}$. Note that $DX = BX - BD = BC\cos B - 2BP\cos B$. Similarly, $EY = EC - CY = 2PC\cos C - BC\cos C$. Hence, we find

$$\frac{DX}{EY} = \frac{BC - 2BP}{2PC - BC}\frac{\cos B}{\cos C} = \frac{\cos B}{\cos C}.$$

Putting this together with our previous computation yields $\frac{DX}{EY} = \frac{XH}{YH}$, as desired.

Solution 2. We adopt the notations of the previous solution and operate on the same assumptions on the configuration. We give an alternate way to see $\frac{DX}{EY} = \frac{XH}{YH}$. Let U_B and U_C be the feet of the perpendiculars from M to AB and AC, respectively. Let V_B and V_C be the feet of the perpendiculars from P to AB and AC, respectively. Note that

$$\frac{DX}{EY} = \frac{2U_BV_B}{2U_CV_C} = \frac{U_BV_B}{U_CV_C}.$$

Let S be the perpendicular of P to U_BM, and let T be the foot of the perpendicular from M to PV_C. Suppose that PS and MT intersect at R. Note that triangles RPM and ABC are similar. Points S and T in RPM correspond to points X and Y in ABC. Hence, we see that

$$\frac{U_BV_B}{U_CV_C} = \frac{PS}{TM} = \frac{BX}{CY} = \frac{HX}{HY}.$$

■

Comment. The result holds if the variable points D and E are chosen such that there exists constants α, β and c for which $\alpha \cdot AD + \beta \cdot AE = c$. The proof in this case is left as an exercise to the reader. On the actual test, some students misread the definition of D and E to be such that $BD = BP$ and $CE = CP$, while others read it to be $BD = DP$ and $CE = EP$. The generalizations aforementioned encompasses all three possible scenarios.

Problem 22. (Cezar Cosnita, Gazeta Matematica) Let ABC be a triangle with circumcenter O and let X, Y, Z be the circumcenters of triangles BOC, COA, AOB. Prove that the lines AX, BY, CZ are concurrent.

Solution 1. We could cheat a little and use the powerful Sondat's theorem (Problem 100) to destroy this problem. Indeed, notice that the perpendiculars from X to BC, from Y to CA, from Z to AB are concurrent at O, since these line are actually the perpendicular bisectors of the sides. On the other hand, the perpendiculars from A to YZ, from B to ZX, from C to XY are also concurrent at O, since we are dealing with the line bisectors of segments AO, BO, and CO, respectively. This implies that triangles ABC and XYZ are orthologic with coincident orthology centers, hence they are perspective.

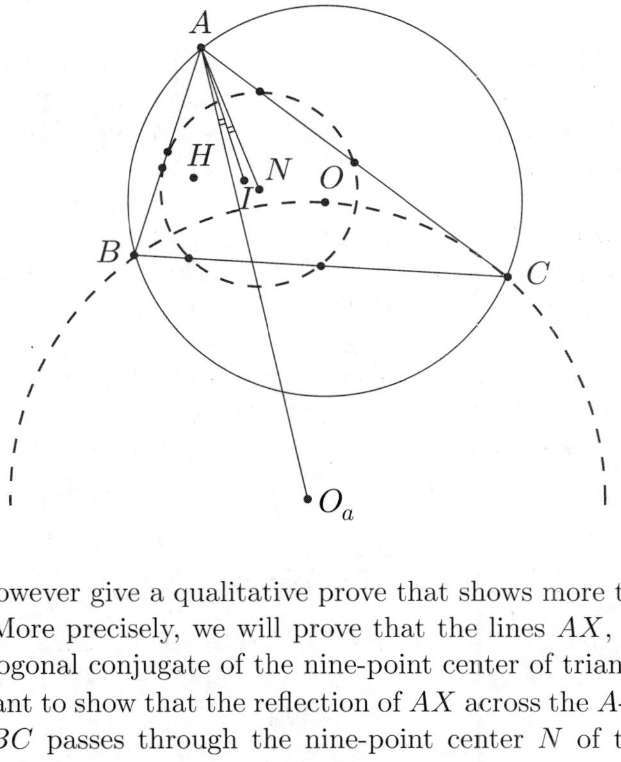

We shall however give a qualitative prove that shows more than the problem asks us. More precisely, we will prove that the lines AX, BY, CZ pass through the isogonal conjugate of the nine-point center of triangle ABC

We thus want to show that the reflection of AX across the A-internal angle bisector of ABC passes through the nine-point center N of triangle ABC. Moreover, since the points O and H are isogonal conjugates with respect to ABC and N is the midpoint of OH, this is equivalent to showing that AX is the A-symmedian of triangle AHO. To see this, we proceed as follows. We work under the assumption that ABC is acute-angled (the solution for the obtuse-angled case will be similar, modulo some diferent angle equalities).

Let P be the foot of the perpendicular from X to OC. Since X is the circumcenter of triangle BOC, P is the midpoint of OC; hence $OP = \frac{R}{2}$, where R is the circumcenter of triangle ABC. Since $\angle COX = \angle A$, we get that

$$OX = \frac{R}{2} \cdot \frac{1}{\cos A}.$$

Let AX meet OH at S. Since $AH \| OX$, it thus follows that

$$\frac{SH}{SO} = \frac{AH}{OX} = \frac{2R\cos A}{\frac{R}{2\cos A}} = 4\cos^2 A.$$

However,

$$4\cos^2 A = \left(\frac{2R\cos A}{R}\right)^2 = \left(\frac{AH}{AO}\right)^2.$$

Hence, we conclude that $\frac{SH}{SO} = \left(\frac{AH}{AO}\right)^2$, which means that AS is the A-symmedian of triangle AHO, as claimed.

Solution 2. This solution also feels a bit like cheating, since we shall use a different result that more or less trivializes the problem. However, its application here is quite surprising, so we could not resit presenting it. The key idea is to use the beautiful

Jacobi's Theorem. Let ABC be a triangle, and let X, Y, Z be three points in its plane such that $\angle YAC = \angle BAZ$, $\angle ZBA = \angle CBX$ and $\angle XCB = \angle ACY$. Then, the lines AX, BY, CZ are concurrent.

Proof. We use directed angles taken modulo 180°. Denote by A, B, C, x, y, z the magnitudes of the angles $\angle CAB$, $\angle ABC$, $\angle BCA$, $\angle YAC$, $\angle ZBA$, and $\angle XCB$, respectively. Since the lines AX, BX, CX are trivially concurrent (at X), the trigonometric version of Ceva's theorem yields

$$\frac{\sin\angle CAX}{\sin\angle XAB} \cdot \frac{\sin\angle ABX}{\sin\angle XBC} \cdot \frac{\sin\angle BCX}{\sin\angle XCA} = 1.$$

We now notice that $\angle ABX = \angle ABC + \angle CBX = B + y$, $\angle XBC = -\angle CBX = -y$, $\angle BCX = -\angle XCB = -z$, $\angle XCA = \angle XCB + \angle BCA = z + C$. Hence, we get

$$\frac{\sin\angle CAX}{\sin\angle XAB} \cdot \frac{\sin(B+y)}{\sin(-y)} \cdot \frac{\sin(-z)}{\sin(C+z)} = 1.$$

Similarly, we can find

$$\frac{\sin\angle ABY}{\sin\angle YBC} \cdot \frac{\sin(C+z)}{\sin(-z)} \cdot \frac{\sin(-x)}{\sin(A+x)} = 1,$$

$$\frac{\sin\angle BCZ}{\sin\angle ZCA} \cdot \frac{\sin(A+x)}{\sin(-x)} \cdot \frac{\sin(-y)}{\sin(B+y)} = 1.$$

Multiplying all these three equations and canceling the same terms, we get

$$\frac{\sin \angle CAX}{\sin \angle XAB} \cdot \frac{\sin \angle ABY}{\sin \angle YBC} \cdot \frac{\sin \angle BCZ}{\sin \angle ZCA} = 1.$$

Therefore, by trigonometric Ceva, the lines AX, BY, CZ are concurrent. This completes the proof of Jacobi's theorem.

□

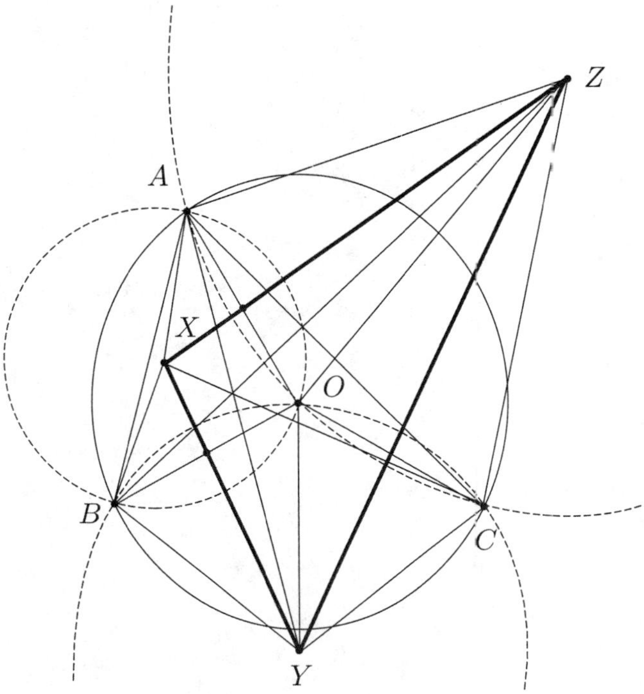

The plan is now to apply Jacobi's theorem for triangle XYZ with outer points A, B and C. Since $ZBOA$ is a kite, we have $\angle BZO = \angle AZO$. But $ZBXO$ and $ZOYA$ are also kites. Therefore, $\angle BZE = \frac{1}{2}\angle BZO$ and $\angle AZY = \frac{1}{2}\angle AZO$, from which it follows that $\angle BZX = \angle AZY$. In a similar fashion it can be shown that $\angle BXZ = \angle CXY$ and $\angle CYX = \angle AYZ$, so Jacobi's theorem yields that lines XA, YB, ZC are concurrent, as desired. This completes the second proof.

■

Problem 23. (Brocard's Theorem) Let ABC be a triangle and let P be a point in its interior it. Let the line passing through P, which is perpendicular to PA, intersect BC at A_1. Analogously, define B_1 and C_1. Prove that A_1, B_1, C_1 are collinear.

Solution. Consider a circle Γ centered at P, with arbitrary radius, and denote the polars of points A, B, C, A_1, B_1, C_1 with respect to Γ by a, b, c, a_1, b_1, c_1, respectively.

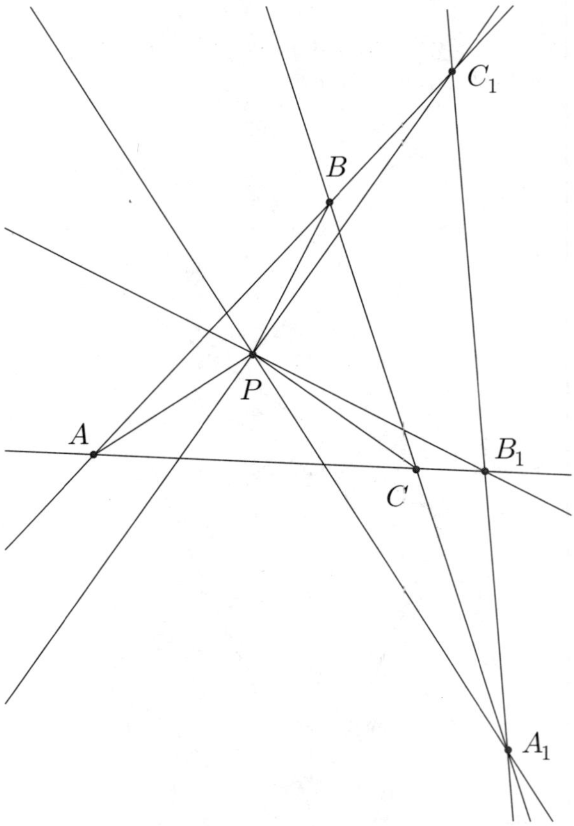

By definition, we have $a \perp PA$, $b \perp PB$, $c \perp PC$, $a_1 \perp PA_1$, $b_1 \perp PB_1$ and $c_1 \perp PC_1$. Furthermore, since the point A_1 lies on BC, the polar a_1 passes through the intersection $b \cap c$. From $a_1 \perp PA_1$ and $PA_1 \perp PA$, we have $a_1 \| PA$; from $a \perp PA$, we thus get $a_1 \perp a$. Hence, a_1 is the line passing through the intersection $b \cap c$ and orthogonal to a. Analogously, b_1 is the line passing through the intersection $c \cap a$ and orthogonal to b, and c_1 is the line passing through the intersection $a \cap b$ and orthogonal to c. Therefore, a_1, b_1, c_1 are the altitudes of the triangle formed by the lines a, b and c. Consequently, the lines a_1, b_1, c_1 concur. From this, we derive that the points A_1, B_1, C_1 are collinear.

■

Comment. The line passing through A_1, B_1, C_1 is called the *orthopolar* of P with respect to triangle ABC and has numerous properties. We prove one beautiful theorem due to Luis Gonzalez.

Orthopolar-Polar Theorem. In addition to the notations already established, let A', B', C' be the intersections of AP with BC, of BP with CA, and of CP with AB, respectively. Also, let X, Y, Z be the projections of P on $B'C'$, $C'A'$, and $A'B'$, respectively. If τ denotes the orthopolar of P with respect to ABC, then τ is the polar of P with respect to the circumcircle of triangle XYZ.

Proof. Let the perpendicular to PA at P cut $A'B'$ and $A'C'$ at N and M, respectively. Let U be the intersection of MN and YZ and let A_2 be the reflection of P across U. Notice that $\angle MNZ = \angle ZPA' = \angle ZYA'$, so $MNZY$ is cyclic, and since UP is clearly tangent to the circumcircle of $PYA'Z$, we have that $UA_2{}^2 = UP^2 = UY \cdot UZ = UM \cdot UN$; consequently, the circle (U) with diameter PA_2 is orthogonal to both circumcircles of XYZ and $MNZY$; this means that points P and A_2 are conjugate with respect to (XYZ). It also follows that (P, A_2, M, N) is a harmonic division; however, $(A'P, A'A_1, A'C', A'B')$ is a harmonic pencil, and thus so is $(A'P, A'A_2, A'C', A'B')$. Hence, we conclude that $A'A_1 \equiv A'A_2 \equiv BC$, which gives $A_1 \equiv A_2$. This means that A_1 lies on the polar of P with respect to the circumcircle of triangle XYZ. Analogously, B_1 and C_1 lie on the polar of P with respect to (XYZ) and so τ is the polar of P with respect to the circumcircle of triangle XYZ. This proves the theorem.

Problem 24. (International Zhautykov Olympiad 2009) Given a quadrilateral $ABCD$ with $\angle B = \angle D = 90^\circ$. Point M is chosen on segment AB so that $AD = AM$. Rays DM and CB intersect at point N. Points H and K are the feet of the perpendiculars from D and C to lines AC and AN, respectively. Prove that $\angle MHN = \angle MCK$.

Solution. First, note that

$$\begin{aligned}\angle NDC &= 90^\circ - \angle ADM\\ &= 90^\circ - \left(90 - \frac{1}{2}\angle BAD\right)\\ &= \frac{1}{2}\angle BAD.\end{aligned}$$

Also,

$$\begin{aligned}\angle DNC &= 180^\circ - \angle BCD - \angle NDC\\ &= 180^\circ - (180^\circ - \angle BAD) - \frac{1}{2}\angle BAD\\ &= \frac{1}{2}\angle BAD.\end{aligned}$$

Hence, it follows that $\angle NDC = \angle DNC$, so $CD = CN$.

Second, note that $AM^2 = AD^2 = AH \cdot AC$, which means that AB is tangent to the circumcircle of triangle HMC. In particular, it follows that $\angle BMC = \angle MHC$. Similarly, $CN^2 = CD^2 = CH \cdot CA$; hence, CN is tangent to the circumcircle of triangle AHN, which gives $\angle CHN = \angle CNA$. Combining the two angle equalities obtained, we conclude that

$$\begin{aligned} \angle MHN &= \angle CHN - \angle MHC \\ &= \angle CNA - \angle BMC \\ &= 90^\circ - \angle KCN - \angle BMC \\ &= \angle MCK. \end{aligned}$$

This completes the proof.

■

Problem 25. (IMO Shortlist 1998) Let P be a point in the plane of a triangle ABC and let Q be its isogonal conjugate with respect to ABC. Prove that

$$\frac{AP \cdot AQ}{AB \cdot AC} + \frac{BP \cdot BQ}{BA \cdot BC} + \frac{CP \cdot CQ}{CA \cdot CB} = 1.$$

Solution. Note that

$$\frac{[AZQY]}{[ABC]} + \frac{[BZQX]}{[ABC]} + \frac{[CXQY]}{[ABC]} = 1,$$

as $[AZQY] + [BZQX] + [CXQY] = [ABC]$. By Lemma 1 from the solution of Problem 18, we know that $AQ \perp YZ$, $BQ \perp ZX$, $CQ \perp XY$; thus, we can write

$$[AZQY] = \frac{1}{2} AQ \cdot YZ, \quad [BZQX] = \frac{1}{2} BQ \cdot ZX \quad \text{and} \quad [CXQY] = \frac{1}{2} CQ \cdot XY.$$

It follows that

$$\begin{aligned} 1 &= \frac{[AZQY]}{[ABC]} + \frac{[BZQX]}{[ABC]} + \frac{[CXQY]}{[ABC]} \\ &= \frac{AQ \cdot YZ}{AB \cdot AC \sin A} + \frac{BQ \cdot ZX}{BC \cdot BA \sin B} + \frac{CQ \cdot XY}{CA \cdot CB \sin C} \\ &= \frac{AP \cdot AQ}{AB \cdot AC} + \frac{BP \cdot BQ}{BA \cdot BC} + \frac{CP \cdot CQ}{CA \cdot CB}, \end{aligned}$$

as desired. The last equality holds because we know that $YZ = AP \sin A$, $ZX = BP \sin B$, $XY = CP \sin C$. This completes the proof.

■

Problem 26. (Lev Emelyanov, Tuymaada Yacut Olympiad 2005) In a triangle ABC, let A_1, B_1, C_1 be the points where the excircles touch the sides BC, CA and AB respectively. Prove that AA_1, BB_1 and CC_1 are the sidelenghts of a triangle.

Solution. The straightforward solution is of course computational. One can confidently compute the precise lengths of segments AA_1, BB_1, CC_1 in terms of the sides of triangle ABC, using barycentric coordinates or Stewart's theorem, afterwhich the problem just turns into algebraic manipulations.

However, we present an argument "from the book". Consider the line through A parallel to BC, the line through B parallel to CA and the line through C parallel to AB. Let $A'B'C'$ be the triangle thus obtained - this is usually called the *anticomplementary triangle* of ABC (because ABC is the medial triangle of $A'B'C'$). Furthermore, let D, E, F be the tangency points of the incircle of ABC with BC, CA, AB respectively. Since the tangency points of the excircles with the sides are just the reflections of D, E, F with respect to the midpoints of BC, CA, AB, respectively, we notice that $A'D = AA_1$, $B'E = BB_1$ and $C'F = CC_1$. Now, fold the triangles $A'BC$, $B'CA$, $C'AB$, which are congruent to ABC, to form a tetrahedron whose base is ABC in such a way that A', B' and C' become the same point, say P. The previous equalities become $PD = AA_1$, $PE = BB_1$, and $PF = CC_1$, respectively. By the triangle inequality, $EF + PE > PF$; on the other hand, however, $PD > EF$, since EF is a chord of the incircle, while PD is at least the length of an altitude of ABC. Hence,

$$PD + PE > EF + PE > PF.$$

Analogously, we get that $PE + PF > PD$ and $PF + PD > PE$. This proves that the lengths $PD = AA_1$, $PE = BB_1$, $PF = CC_1$ can be the side lengths of a triangle, as claimed.

■

Problem 27. (Morley's Theorem) Prove that the three points of intersections of the adjacent internal angle trisectors of a triangle form an equilateral triangle.

Solution 1. There are many ways of thinking about this - ways that are conceptually very different. We will cover three of these proofs. The first proof, the less sophisticated one, is a straightforward attempt to express the side of the so-called Morley triangle in terms of the sides of ABC, with the hope of getting something symmetric in a, b, c at the end. While convoluted and boring, the method does indeed work and is very instructive, so we shall cover it in full detail below.

We want to show that the triangle $E_1E_2E_3$ is equilateral. Let $\mathcal{R}$ denote the circumradius of $A_1A_2A_3$. Setting $\angle A_i = 3\theta_i$ for $i = 1, 2, 3$, we get $\theta_1+\theta_2+\theta_3 = 60^\circ$. We now apply the Law of Sines twice to deduce

$$\begin{aligned} A_1E_3 &= \frac{\sin\theta_2}{\sin(180^\circ - \theta_1 - \theta_2)} A_1A_2 \\ &= \frac{\sin\theta_2}{\sin(120^\circ + \theta_3)} \cdot 2\mathcal{R}\sin(3\theta_3) \\ &= 8\mathcal{R}\sin\theta_2\sin\theta_3\sin(60^\circ + \theta_3), \end{aligned}$$

where the latter equality holds because of the well-known formula

$$\sin(3\theta) = 4\sin\theta\sin(60^\circ + \theta)\sin(120^\circ + \theta).$$

By symmetry, we also have

$$A_1E_2 = 8\mathcal{R}\sin\theta_3\sin\theta_2\sin(60^\circ + \theta_2).$$

Now, we present two different ways to complete the proof. The first method is more direct, whereas the second method is nicer because it gives more information.

First Approach. We apply the Law of Cosines to obtain

$$\begin{aligned} & E_1E_2{}^2 \\ = \; & AE_3{}^2 + AE_2{}^2 - 2\cos(\angle E_3A_1E_2)\cdot AE_3\cdot AE_1 \\ = \; & 64\mathcal{R}^2\sin^2\theta_2\sin^2\theta_3[\sin^2(60^\circ+\theta_3) + \sin^2(60^\circ+\theta_2) \\ - \; & 2\cos\theta_1\sin(60^\circ+\theta_3)\sin(60^\circ+\theta_2)] \end{aligned}$$

To avoid long computation, here, we employ a trick. Namely, we shall use the fact that all A, B, C with $A + B + C = 180^\circ$, we have

$$\cos^2 A + \cos^2 B + \cos^2 C + 2\cos A\cos B\cos C = 1.$$

In view of the equality

$$(180^\circ - \theta_1) + (30^\circ - \theta_2) + (30^\circ - \theta_3) = 180^\circ,$$

we thus have

$$\cos^2(180^\circ - \theta_1) + \cos^2(30^\circ - \theta_2) + \cos^2(30^\circ - \theta_3)$$

$$+2\cos(180^\circ - \theta_1)\cos(30^\circ - \theta_2)\cos(30^\circ - \theta_3) = 1$$

or

$$\cos^2\theta_1+\sin^2(60^\circ+\theta_2)+\sin^2(60^\circ+\theta_3)-2\cos\theta_1\sin(60^\circ+\theta_2)\sin(60^\circ+\theta_3)=1$$

or

$$\sin^2(60^\circ+\theta_2)+\sin^2(60^\circ+\theta_3)-2\cos\theta_1\sin(60^\circ+\theta_2)\sin(60^\circ+\theta_3)=\sin^2\theta_1.$$

We therefore find that

$$E_1E_2{}^2=64\mathcal{R}^2\sin^2\theta_1\sin^2\theta_2\sin^2\theta_3$$

so that

$$E_1E_2=8\mathcal{R}\sin\theta_1\sin\theta_2\sin\theta_3$$

Remarkably, the length of E_1E_2 is symmetric in the angles! By symmetry, we therefore conclude that $E_1E_2E_3$ is an equilateral triangle with the length $8\mathcal{R}\sin\theta_1\sin\theta_2\sin\theta_3$. This completes the first approach.

Second Approach. We find the angles in the picture explicitly. Look at the triangle $E_3A_1E_2$. The equality

$$\theta_1+(60^\circ+\theta_2)+(60^\circ+\theta_3)=180^\circ$$

allows us to invite a ghost triangle ABC having the angles

$$A=\theta_1,\ B=60^\circ+\theta_2,\ C=60^\circ+\theta_3.$$

Observe that two triangles BAC and $E_3A_1E_2$ are similar. Indeed, we have $\angle BAC=\angle E_3A_1E_2$ and

$$\frac{A_1E_3}{A_1E_2}=\frac{8\mathcal{R}\sin\theta_2\sin\theta_3\sin(60^\circ+\theta_3)}{8\mathcal{R}\sin\theta_3\sin\theta_2\sin(60^\circ+\theta_2)}=\frac{\sin(60^\circ+\theta_3)}{\sin(60^\circ+\theta_2)}=\frac{\sin C}{\sin B}=\frac{AB}{AC}.$$

It therefore follows that

$$(\angle A_1E_3E_2,\ \angle A_1E_2E_3)=(60^\circ+\theta_2,\ 60^\circ+\theta_3).$$

Similarly, we also have

$$(\angle A_2E_1E_3,\ \angle A_2E_3E_1)=(60^\circ+\theta_3,\ 60^\circ+\theta_1).$$

and

$$(\angle A_3E_2E_1,\ \angle A_3E_1E_2)=(60^\circ+\theta_1,\ 60^\circ+\theta_2).$$

An angle computation yields

$$\begin{aligned}\angle E_1E_2E_3 &= 360^\circ - (\angle A_1E_2E_3 + \angle E_1E_2A_3 + \angle A_3E_2A_1) \\ &= 360^\circ - [\,(60^\circ + \theta_3) + (60^\circ + \theta_1) + (180^\circ - \theta_3 - \theta_1)\,] \\ &= 60^\circ.\end{aligned}$$

Similarly, we also have $\angle E_2E_3E_1 = 60^\circ = \angle E_3E_1E_2$. It follows that $E_1E_2E_3$ is equilateral. Furthermore, we apply the Law of Sines to reach

$$\begin{aligned}E_2E_3 &= \frac{\sin\theta_1}{\sin(60^\circ + \theta_3)}\,A_1E_3 \\ &= \frac{\sin\theta_1}{\sin(60^\circ + \theta_3)} \cdot 8\mathcal{R}\sin\theta_2\sin\theta_3\sin(60^\circ + \theta_3) \\ &= 8\mathcal{R}\sin\theta_1\sin\theta_2\sin\theta_3.\end{aligned}$$

Hence, we find that the triangle $E_1E_2E_3$ has the length $8\mathcal{R}\sin\theta_1\sin\theta_2\sin\theta_3$.

■

Solution 2. The second proof we give is inarguably the simplest proof known. Let our triangle have angles $3a$, $3b$, $3c$ and let x^* mean $x + 60^\circ$, so that $a + b + c = 0^*$. Then triangles with angles $(0^*, 0^*, 0^*)$, (a, b^*, c^*), (a^*, b, c^*), (a^*, b^*, c), (a^{**}, b, c), (a, b^{**}, c), (a, b, c^{**}) exist abstractly, since in every case the angle-sum is 180°. Build them on a scale defined as follows:

- $(0^*, 0^*, 0^*)$ - this is equilateral - make it have edge 1;

- (a, b^*, c^*) - make the edge joining the angles b^* and c^* have length 1 (and similarly for (a^*, b, c^*), (a^*, b^*, c));

- (a^{**}, b, c) (and the other two like it) as follows: let the angles at B, P, C be b, a^{**}, c, and draw lines from P cutting BC at angle a^* in the two senses, so forming an isosceles triangle PYZ. Choose the scale so that PY and PZ are both 1.

Now just fit all these 7 triangles together! They'll form a figure that looks like

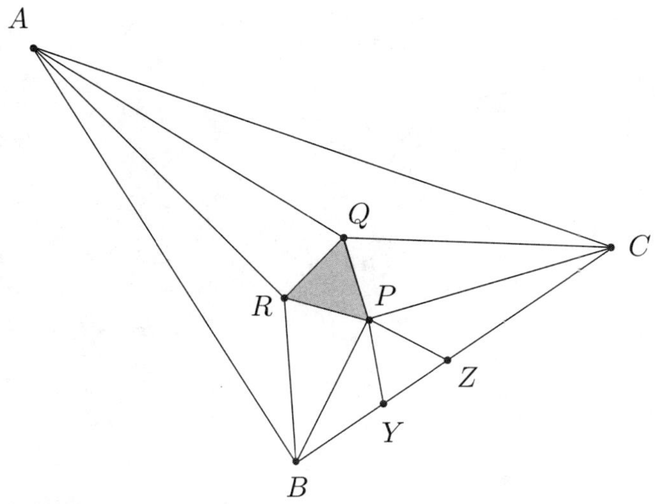

(in which the points X, Y should really be omitted). To make it a bit clearer, let us say that the angles of triangle BPR are b (at B), c^* (at P), a^* (at R). Why do they all fit together? Well, at each internal vertex, the angles add up to $360°$, as one can easily check. And two coincident edges have either both been declared to have length 1, or are like the common edge BP of triangles BPR and BPC. But triangle BPR is congruent to the subtriangle BPZ of triangle BPC, since $PR = PZ = 1$, $\angle PBR = \angle PBZ = b$, and $\angle BRP = \angle BZP = a°$.

So the figure formed by these 7 triangles is similar to the one we get by trisecting the angles of the given triangle, and therefore in that triangle the middle subtriangle must also be equilateral. This completes the second proof of Morley's Theorem.

■

Solution 3. The third proof is a really beautiful conceptual by Alain Connes and it comes as a group theoretic property of the action of the affine group on the line. It will be valid for any (commutative) field k (with arbitrary characteristic, though in characteristic 3 the hypothesis of the theorem cannot be fulfilled). Thus, we let k be such a field and let G be the affine group over K, in other words the group of 2×2 matrices $g = \begin{pmatrix} a & b \\ 0 & 1 \end{pmatrix}$ where $a \in k$, $a \neq 0$, $b \in k$. For $g \in G$ we let,

$$\delta(g) = a \in k^*.$$

By construction δ is a morphism from G to the multiplicative group k^* of non zero elements of k, and the subgroup $T = \mathrm{Ker}\delta$ is the group of translations, i.e., the additive group of k. Each $g \in G$ defines a transformation,

$$g(x) = ax + b, \quad \text{for all } x \in k,$$

and if $a \neq 1$, it admits one and only one fixed point,

$$\mathrm{fix}(g) = \frac{b}{1-a}.$$

Let us prove the following simple fact:

Theorem. Let g_1, g_2, g_3 be in G such that g_1g_2, g_2g_3, g_3g_1 and $g_1g_2g_3$ are not translations and let $j = \delta(g_1g_2g_3)$. The following two conditions are equivalent,

a) $g_1^3g_2^3g_3^3 = 1$.

b) $j^3 = 1$ and $\alpha + j\beta + j^2\gamma = 0$ where $\alpha = \mathrm{fix}(g_1g_2)$, $\beta = \mathrm{fix}(g_2g_3)$, $\gamma = \mathrm{fix}(g_3g_1)$.

Proof of Theorem. Let $g_i = \begin{pmatrix} a_i & b_i \\ 0 & 1 \end{pmatrix}$. The equality $g_1^3g_2^3g_3^3 = 1$ is equivalent to $\delta(g_1^3g_2^3g_3^3) = 1$, and $b = 0$, where b is the translational part of $g_1^3g_2^3g_3^3$. The first condition is exactly $j^3 = 1$. Note that $j \neq 1$ by hypothesis. Next one has

$$b = (a_1^2 + a_1 + 1)b_1 + a_1^3(a_2^2 + a_2 + 1)b_2 + (a_1a_2)^3(a_3^2 + a_3 + 1)b_3.$$

A straightforward computation, using $a_1a_2a_3 = j$ gives,

$$b = -ja_1^2a_2(a_1 - j)(a_2 - j)(a_3 - j)(\alpha + j\beta + j^2\gamma),$$

where α, β, γ are the fixed points

$$\alpha = \frac{a_1b_2 + b_1}{1 - a_1a_2}, \quad \beta = \frac{a_2b_3 + b_2}{1 - a_2a_3}, \quad \gamma = \frac{a_3b_1 + b_3}{1 - a_3a_1}.$$

Now, $a_k - j \neq 0$ since by hypothesis the pairwise products of g_j's are not translations. Thus, and whatever the characteristic of k is, we get that a) is equivalent with b).

Returning to Morley's theorem and its proof, take $k = \mathbb{C}$ and let g_1 be the rotation with center A and angle $2a$, where a is the angle BAC and similarly for g_2, g_3. One has $g_1^3g_2^3g_3^3 = 1$ since each g_i^3 can be expressed as the product of the symmetries along the consecutive sides. Moreover for a similar reason $\alpha = \mathrm{fix}(g_1g_2)$, $\beta = \mathrm{fix}(g_2g_3)$, $\gamma = \mathrm{fix}(g_3g_1)$ are the intersections of trisectors. Thus from a) $\Longrightarrow$ b), one gets $\alpha + j\beta + j^2\gamma = 0$ which is a classical characterization of equilateral triangles. This completes the third proof.

■

Comment. Note that without altering the cubes g_1^3, g_2^3, g_3^3 one can multiply each g_i by a cubic root of 1; one obtains in this way the 18 nondegenerate equilateral triangles of variants of Morley's theorem!

Problem 28. (Nguyen Minh Ha, Mathematics and Youth Magazine) Let Ω be the circumcircle of triangle ABC and let D be the tangency point of its incircle $\rho(I)$ with the side BC. Let ω be the circle internally tangent to Ω at T, and to BC at D. Prove that $\angle ATI = 90°$.

Solution. We begin with a simple Lemma.

Lemma. Let γ be a circle and let A and B be two arbitrary points on it. A circle ρ touches γ internally at T. Denote by AE and BF the tangency lines to ρ at E and F, respectively. Then

$$\frac{TA}{TB} = \frac{AE}{BF}.$$

Proof. Denote by A_1 and B_1 the second intersections of TA and TE with ρ, respectively. We know that A_1B_1 is parallel to AB. Therefore,

$$\left(\frac{AE}{TA_1}\right)^2 = \frac{AA_1 \cdot AT}{A_1T \cdot A_1T} = \frac{BB_1}{B_1T} \cdot \frac{BT}{B_1T} = \left(\frac{BF}{TB_1}\right)^2.$$

Hence,

$$\frac{AE}{TA_1} = \frac{BF}{TB_1},$$

which gives

$$\frac{AE}{BF} = \frac{TA_1}{TB_1} = \frac{TA}{TB},$$

which completes the proof of the Lemma.

Returning to the problem, let E and F be the tangency points of $\rho(I)$ with sides CA and AB, respectively. According to the Lemma,

$$\frac{TB}{TC} = \frac{BD}{CD} = \frac{BF}{CE}.$$

Therefore triangles TBF and TCE are similar. It follows that $\angle TFA = \angle TEA$; hence points A, I, E, F, T all lie on the same circle. We conclude that $\angle ATI = \angle AFI = 90°$, which completes our proof.

■

Problem 29. (Hartcourt's Theorem) Let ABC be a triangle and let ℓ be a tangent to its incircle. Let x, y, z be the signed distances from A, B, C to ℓ, respectively. Prove that

$$ax + by + cz = 2[ABC],$$

where $[ABC]$ denotes the area of triangle ABC.

Solution. We first recall two beautiful results that we shall use in the proof.

Salmon's Theorem. Let Γ be a circle with center O and let P and Q be two points in its plane. Let p and q denote the polars of P and Q with respect to Γ. Then,

$$\frac{OP}{OQ} = \frac{\delta(P,q)}{\delta(Q,p)},$$

where by $\delta(X,\ell)$ we denote the distance from X to the line ℓ.

Euler's Pedal Triangle Theorem. Let P be a point in the plane of the triangle ABC. If A_1, B_1, C_1 are the projections of P on the sides BC, CA, AB, then

$$\frac{[A_1B_1C_1]}{[ABC]} = \frac{|R^2 - OP^2|}{4R^2},$$

where $[XYZ]$ denotes the area of triangle XYZ.

Both results are well known and have quite beautiful synthetic solutions. For sake of completeness, we include them below.

Proof of Salmon's Theorem. We will just take care of the case when both P and Q are outside Γ (the proofs for the other cases can be adapted). In this case, the polars p and q are just the lines determined by the intersection points of the tangents from P and Q with Γ, respectively. Let P' and Q' be the intersections of OP and OQ with p and q. We have that

$$OP \cdot OP' = OQ \cdot OQ' = r^2;$$

therefore

$$\frac{OP}{OQ} = \frac{OQ'}{OP'}.$$

We are thus left to show that

$$\frac{OQ'}{OP'} = \frac{\delta(P,q)}{\delta(Q,p)}.$$

But this is immediate. Let the orthogonal projections from P and Q to the lines q and p be X and Y, respectively. We have that $OP \perp p$ and $OQ \perp q$, so quadrilaterals $PXQ'O$ and $QYP'O$ are similar. Hence

$$\frac{OQ'}{OP'} = \frac{PX}{QY},$$

as desired.

□

Proof of Euler's Pedal Triangle Theorem. This is slightly more involved. We need a preliminary result that will provide a lot of insight into what is about to happen.

Claim. Let X, Y, Z be the intersections of the lines AP, BP, CP with the circumcircle of ABC, other than the vertices of the triangle. Then, triangles $A_1B_1C_1$ and XYZ are similar.

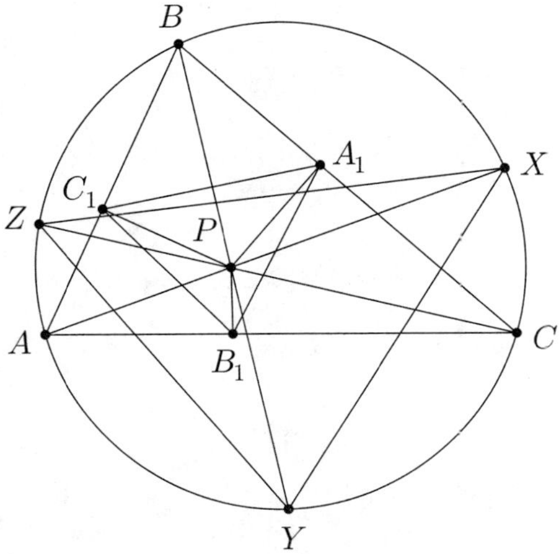

Indeed, we can write

$$\begin{aligned}\angle B_1A_1C_1 &= \angle PA_1B_1 + \angle PA_1C_1 \\ &= \angle PCB_1 + \angle PBC_1 \\ &= \angle ZCA + \angle YBA \\ &= \angle ZXA + \angle YXA \\ &= \angle YXZ.\end{aligned}$$

And similarly, we can show that $\angle B_1C_1A_1 = \angle YZX$ and $\angle A_1B_1C_1 = \angle XYZ$, so the claim is indeed true.

Returning to the proof, we start computing the ratio $\frac{[A_1B_1C_1]}{[ABC]}$ using the well known formula $YZ \cdot ZX \cdot XY = 4R_{XYZ} \cdot [XYZ]$, where R_{XYZ} denotes the circumradius of triangle ABC. Consequently, we write

$$\frac{[A_1B_1C_1]}{[ABC]} = \frac{B_1C_1}{BC} \cdot \frac{C_1A_1}{CA} \cdot \frac{A_1B_1}{AB} \cdot \frac{R}{R_{A_1B_1C_1}}.$$

However, triangle $A_1B_1C_1$ and the circumcevian triangle XYZ are similar, and moreover $R_{XYZ} = R$, so

$$\frac{R}{R_{A_1B_1C_1}} = \frac{R_{XYZ}}{R_{A_1B_1C_1}} = \frac{YZ}{B_1C_1}.$$

Hence, we get that

$$\frac{[A_1B_1C_1]}{[ABC]} = \frac{YZ}{BC} \cdot \frac{C_1A_1}{CA} \cdot \frac{A_1B_1}{AB}.$$

On the other hand, the quadrilateral $BCYZ$ is cyclic, so triangles PBC and PZY are similar, therefore $\frac{PB}{PZ} = \frac{PC}{PY} = \frac{BC}{YZ}$; thus keeping in mind that $C_1A_1 = PB\sin B$ and $A_1B_1 = PC\sin C$ (which follows from the Law of Sines in triangle BA_1C_1 and CA_1B_1), we can write

$$\begin{aligned}\frac{[A_1B_1C_1]}{[ABC]} &= \frac{YZ}{BC} \cdot \frac{C_1A_1}{CA} \cdot \frac{A_1B_1}{AB} \\ &= \frac{PZ}{PB} \cdot \frac{PB\sin B}{CA} \cdot \frac{PC\sin C}{AB} \\ &= \frac{PZ}{PB} \cdot \frac{PB}{2R} \cdot \frac{PC}{2R} \\ &= \frac{PC \cdot PZ}{4R^2} \\ &= \frac{\left|R^2 - OP^2\right|}{4R^2},\end{aligned}$$

where the last equality holds since $PC \cdot PZ$ is precisely the power of P with respect to the circumcircle of ABC. This completes the proof.

□

Returning to the proof of the original, let D, E, F be the tangency points of the incircle of ABC with the sides BC, CA, AB, respectively.

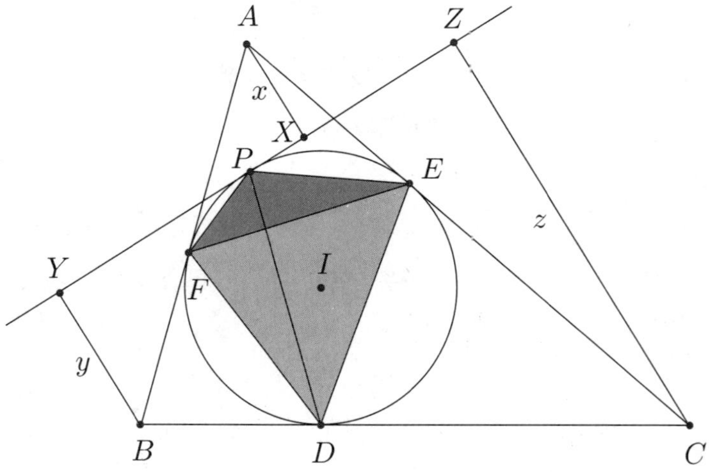

From Euler's pedal triangle theorem, we have that

$$\frac{[DEF]}{[ABC]} = \frac{R^2 - OI^2}{4R^2} = \frac{r}{2R},$$

where the latter equality holds because of the well known Euler identity $OI^2 = R^2 - 2Rr$.

By Salmon's lemma, applied for the tangency point P of ℓ with the incircle and vertex A, we get

$$\frac{IP}{IA} = \frac{\delta(P, EF)}{\delta(A, \ell)}, \quad \text{or equivalently} \quad ax = \frac{a \cdot IA \cdot \delta(P, EF)}{r}.$$

Analogously, we obtain

$$by = \frac{b \cdot IB \cdot \delta(P, FD)}{r} \quad \text{and} \quad cz = \frac{c \cdot IC \cdot \delta(P, DE)}{r}.$$

Suppose now, without loss of generality, that points B and C are on the same side of ℓ. We thus want to show that

$$by + cz - ax = 2[ABC].$$

We use the formulas above. We write that

$$\begin{aligned}
& by + cz - ax \\
= & \frac{b \cdot IB \cdot \delta(P, FD)}{r} + \frac{c \cdot IC \cdot \delta(P, DE)}{r} - \frac{a \cdot IA \cdot \delta(P, EF)}{r} \\
= & \frac{1}{r}(2R \cdot IB \sin B \cdot \delta(P, FD) + 2R \cdot IC \sin C \cdot \delta(P, DE) \\
- & 2R \cdot IA \sin A \cdot \delta(P, EF)) \\
= & \frac{1}{r}(2R \cdot FD \cdot \delta(P, FD) + 2R \cdot DE \cdot \delta(P, DE) - 2R \cdot EF \cdot \delta(P, EF)) \\
= & \frac{4R}{r}([PFD] + [PDE] - [PEF]) \\
= & \frac{4R}{r}[DEF] \\
= & \frac{4R}{r} \cdot \frac{r}{2R} \cdot [ABC] \\
= & 2[ABC],
\end{aligned}$$

as desired. This completes the proof.

■

Comment. The result we just proved is known as *Hartcourt's theorem*. There is also a different proof using homogeneous barycentric coordinates which is possible, but which we will leave out. However, we remark that the result can be extended to excircles (and invite the reader to find a proof for the following statement).

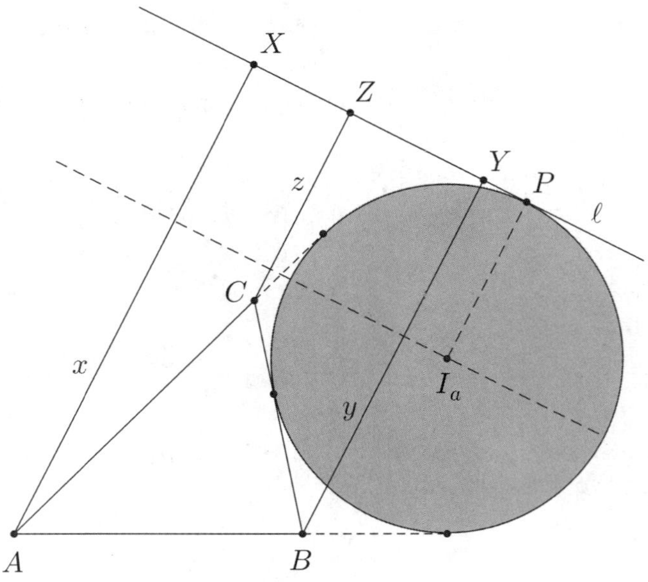

Hartcourt's Ex-Theorem. If the distances from the vertices A, B, C to a tangent to the A-excircle of triangle ABC are x, y, z, respectively, then

$$-ax + by + cz = 2[ABC].$$

Analogous statements hold for the B- and C-excircles.

Problem 30. (Neuberg-Pedoe Inequality) Let a, b, c, and x, y, z be the side lengths of two given triangles ABC, XYZ with areas $[ABC]$, and $[XYZ]$, respectively. Then,

$$a^2\left(y^2+z^2-x^2\right)+b^2\left(z^2+x^2-y^2\right)+c^2\left(x^2+y^2-z^2\right) \geq 16[ABC][XYZ],$$

with equality if and only if the triangles ABC and XYZ are similar.

Solution. First note that the inequality is homogeneous in the side lengths x, y, z of the triangle XYZ (in fact, these side lengths are squared on both sides, since the area of a triangle is quadratically dependent on its side lengths). Hence, this inequality is invariant under any similitude transformation executed on triangle XYZ. In other words, we can move, reflect, rotate and stretch the triangle XYZ as we wish, but the inequality remains equivalent. But, of course, by applying similitude transformations to triangle XYZ, we can always achieve a situation when $Y = B$ and $Z = C$ and the point X lies in the same half-plane with respect to the line BC as the point A. Hence, in order to prove the inequality for any two triangles ABC and XYZ, it is enough to prove it for two triangles ABC and XYZ in this special situation.

Accordingly, assume that the triangles ABC and XYZ are in this special situation, namely $Y = B$, $Z = C$ and the point X lies in the same half-plane with respect to the line BC as the point A. We thus have to prove the inequality

$$a^2\left(y^2+z^2-x^2\right)+b^2\left(z^2+x^2-y^2\right)+c^2\left(x^2+y^2-z^2\right)\geq 16[ABC][XYZ].$$

Well, by the cosine law in triangle ABX, we have

$$AX^2 = AB^2 + XB^2 - 2\cdot AB\cdot XB\cdot\cos\angle ABX.$$

Let's figure out now what this equation means. At first, $AB = c$. Then, since $B = Y$, we have $XB = XY = z$. Finally, we have either $\angle ABX = \angle ABC - \angle XBC$ or $\angle ABX = \angle XBC - \angle ABC$ (depending on the arrangement of the points), but in both cases $\cos\angle ABX = \cos(\angle ABC - \angle XBC)$. Since $B = Y$ and $C = Z$, we can rewrite the angle $\angle XBC$ as $\angle XYZ$. Thus,

$$\begin{aligned}\cos\angle ABX &= \cos\left(\angle ABC - \angle XYZ\right)\\ &= \cos\angle ABC\cos\angle XYZ + \sin\angle ABC\sin\angle XYZ.\end{aligned}$$

By the Law of Cosines in triangles ABC and XYZ, we have

$$\cos\angle ABC = \frac{c^2+a^2-b^2}{2ca}\quad\text{and}\quad\cos\angle XYZ = \frac{z^2+x^2-y^2}{2zx}.$$

Also, since

$$\sin\angle ABC = \frac{2[ABC]}{ca}\quad\text{and}\quad\sin\angle XYZ = \frac{2[XYZ]}{zx},$$

we have that

$$\begin{aligned}&\cos\angle ABX\\ =\;&\cos\angle ABC\cos\angle XYZ + \sin\angle ABC\sin\angle XYZ\\ =\;&\frac{c^2+a^2-b^2}{2ca}\cdot\frac{z^2+x^2-y^2}{2zx}+\frac{2[ABC]}{ca}\cdot\frac{2[XYZ]}{zx}.\end{aligned}$$

This makes the equation

$$AX^2 = AB^2 + XB^2 - 2\cdot AB\cdot XB\cdot\cos\angle ABX$$

transform into

$$AX^2 = c^2+z^2-2cz\left(\frac{c^2+a^2-b^2}{2ca}\cdot\frac{z^2+x^2-y^2}{2zx}+\frac{2[ABC]}{ca}\cdot\frac{2[XYZ]}{zx}\right),$$

which immediately simplifies to

$$AX^2 = c^2 + z^2 - 2\left(\frac{(c^2+a^2-b^2)(z^2+x^2-y^2)}{4ax} + \frac{4[ABC][XYZ]}{ax}\right),$$

and since $YZ = BC$,

$$AX^2 = \frac{(a^2(y^2+z^2-x^2)+b^2(z^2+x^2-y^2)+c^2(x^2+y^2-z^2))-16[ABC][XYZ]}{2ax}.$$

Thus, according to the (obvious) fact that $AX^2 \geq 0$, we conclude that

$$a^2(y^2+z^2-x^2)+b^2(z^2+x^2-y^2)+c^2(x^2+y^2-z^2) \geq 16[ABC][XYZ],$$

which proves the inequality. The equality holds if and only if the points A and X coincide, i.e., if the triangles ABC and XYZ are congruent. Now, of course, since the triangle XYZ we are dealing with is not the initial triangle XYZ, but just its image under a similitude transformation, the general equality condition is that the triangles ABC and XYZ are similar (not necessarily congruent). ■

Comment. The result is known as the *Neuberg-Pedoe Inequality* and comes as a vast generalization of Weitzenbock's inequality from the solution of Problem 72.

Problem 31. (International Zhautykov Olympiad 2011) Diagonals of a cyclic quadrilateral $ABCD$ intersect at point K. The midpoints of segments AC and BD are M and N, respectively. The circumcircles of triangles ADM and BCM intersect at points M and L. Prove that K, L, M and N all lie on a circle.

Solution. Let DA intersect CB at E and let the circumcircles of triangles ADK and BCK intersect again at F. The points E, M, L are collinear on the radical axis of circles (ADM) and (BCM). Similarly, E, K, F are collinear on the radical axis of (ADK) and (BCK).

Since $EM \cdot EL = EB \cdot EC = EK \cdot EF$, Power of Point yields that quadrilateral $MLFK$ is cyclic. However, triangles FBC and FDA are similar; thus, by spiral similarity, so are triangles FBD and FCA. Hence, since M and N are the midpoints of sides CA and BD, we get

$$\angle NFM = \angle DFA = \angle DKA,$$

so $NFKM$ is cyclic. It follows that $KLNM$ is cyclic, as claimed. ■

Problem 32. (Titu Andreescu and Cosmin Pohoata, USAJMO 2014) Let ABC be a triangle with incenter I, incircle γ and circumcircle Γ. Let M, N, P be the midpoints of sides BC, CA, AB and let E, F be the tangency points of γ with CA and AB, respectively. Let U, V be the intersections of EF with MN and MP, respectively, and let X be the midpoint of arc BAC of Γ. Prove that XI bisects UV.

Solution. The first thing that should come to mind is the following observation.

Claim. The points U and V lie on the bisectors BI and CI, respectively.

This claim is a quite popular result that has appeared many times as a lemma in contests problems (and which will appear again at least one more time in the book). We managed to trace it back to Honsberger's famous "Episodes in Nineteeth and Twentieth Century Euclidean Geometry"; however, it is quite possible for this to have been known centuries before. We shall give a modern proof.

Proof of Claim. Let D be the tangency point of γ with BC and let U' be the intersection of BI with EF. Note that triangles BFU' and BDU' are congruent (by SAS), so $\angle BU'F = \angle BU'D$. In addition, the pencil $(U'F, U'B, U'D, U'C)$ is harmonic; thus, it follows that $U'B \perp U'C$, so, in particular, $U'M = MB$, which gives $\angle MU'B = \angle MBU' = \frac{1}{2}\angle A = \angle ABU'$; thus, $MU' \| AB$; hence $U' = U$, which proves the claim that U lies on BI. Similarly, we get that V is on CI. This proves the claim.

Moreover, note that we proved that $IB \perp CU$ and $IC \perp VB$. Next, we show that the lines XB and XC are tangent to the circumcircle of triangle IBC; indeed, observe that

$$\begin{aligned}
\angle XBI &= \angle ABI - \angle ABX \\
&= \frac{1}{2}\angle B - (\angle BCX - \angle C) \\
&= \frac{1}{2}\angle B - \frac{1}{2}(180^\circ - \angle A) + \angle C \\
&= \frac{1}{2}\angle C \\
&= \angle BCI.
\end{aligned}$$

Similarly, $\angle XCI = \angle IBC$. This means that X is the intersection of the tangents at B and C to the circumcircle of IBC; hence, by a well-known property, IX is the I-symmedian of triangle IBC.

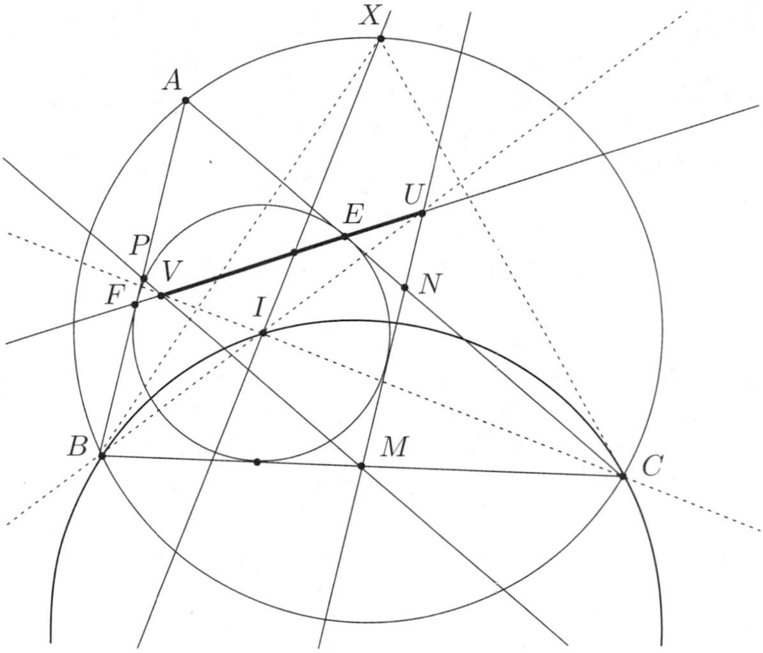

But we proved before that U and V are on IB and IC, respectively, and that $IB \perp CU$ and $IC \perp VB$. In other words, we showed that U and V are the feet of the altitudes from C and B in triangle IBC—so, in particular, we have that $BCUV$ is cyclic and that UV is an antiparallel to BC in triangle IBC. This yields the conclusion, since we know that the I-symmedian of IBC is the locus of the midpoints of the antiparallels to BC in triangle IBC; hence, we showed that IX bisects UV, as claimed.

■

Problem 33. (Cosmin Pohoata, Mathematical Reflections) Let ABC be a triangle and let M, N, P be the midpoints of sides BC, CA, AB, respectively. Denote by X, Y, Z the midpoints of the altitudes emerging from the vertices A, B, C, respectively. Prove that the radical center of the circles AMX, BNY, CPZ is the nine-point center of triangle ABC.

Solution. Let ω_1, ω_2 and ω_3 be the circumcircles of triangles AMX, BNY and CPZ, respectively. Let H be the orthocenter and R be the center of the nine-point circle ω of triangle ABC. Let RM intersect ω_1 at point E and RN intersect ω_2 at point F. Denote by H_1 and H_2 the midpoints of segments AH and BH, respectively. Because EM passes through the center R of ω and it is well known that H_1 is on ω and H_1M is a diameter of ω, it follows that the points E, H_1, R, M lie on a line. In the same way, we have that F, H_2, R, N lie on a line. Denote by U and V the feet of the altitudes from A and B of triangle ABC, respectively. It is obvious that quadrilateral $ABUV$ is cyclic. By the Power of Point theorem, we have

$$\begin{aligned} HA \cdot HU &= HB \cdot HV \\ \left(\frac{HA}{2}\right)\left(\frac{AU}{2} - \frac{HA}{2}\right) &= \left(\frac{HB}{2}\right)\left(\frac{BV}{2} - \frac{HB}{2}\right) \\ AH_1(AX - AH_1) &= BH_2(BY - BH_2) \\ AH_1 \cdot H_1X &= BH_2 \cdot H_2Y \\ EH_1 \cdot H_1M &= FH_2 \cdot H_2N. \end{aligned}$$

But $H_1M = H_2N$ are diameters of ω, therefore, we have that $EH_1 = FH_2$, and so

$$\begin{aligned} RM \cdot RE &= RM(RH_1 + EH_1) \\ &= RN(RH_2 + FH_2) \\ &= RN \cdot RF. \end{aligned}$$

Thus, the powers of R with respect to ω_1 and ω_2 are equal; and since we can do the same thing for the other two pairs of circles, we can hereby conclude that R is the radical center of ω_1, ω_2, ω_3. This completes the proof.

■

Problem 34. (Pavel Kozhevnikov, RMM 2010) Let $A_1A_2A_3A_4$ be a quadrilateral with no pair of parallel sides. For each $i = 1, 2, 3, 4$, define ω_1 to be the circle touching the quadrilateral externally, and which is tangent to the lines $A_{i-1}A_i$, A_iA_{i+1}, and $A_{i+1}A_{i+2}$ (indices are considered modulo 4 so $A_4 = A_0$, $A_5 = A_1$, and $A_6 = A_2$). Let T_i be the point of tangency of ω_i with the side A_iA_{i+1}. Prove that the lines A_1A_2, A_3A_4, and T_2T_4 are concurrent if and only if the lines A_2A_3, A_4A_1, and T_1T_3 are concurrent.

Solution 1. Let A_1A_4 and A_2A_3 intersect at P, and A_1A_2 and A_3A_4 intersect at Q. It is clearly sufficient to show that if P, T_1, and T_3 are collinear then Q, T_4, and T_2 are collinear. Let PT_1 intersect ω_1 at U, let ω_1 be tangent to PA_2 and PA_1 at W and X, respectively, and let ω_4 be tangent to QA_1 and QA_4 at Y and Z, respectively. Because P, T_1, and T_3 are collinear, the homothety centered at P mapping ω_3 to ω_1 maps T_3 to U, so the tangent to ω_1 through U is parallel to A_4A_3. It follows that the homothety mapping ω_1 to ω_4 sends U to Z. Furthermore, we note that this homothety maps T_1 to Y and X to T.

Suppose this homothety also maps W to some point W' on ω_4. Then the tangent to ω_4 through W' is parallel to A_2A_3. Furthermore, because $XUWT_1$ is harmonic, $T_4ZW'Y$ is harmonic, so Q, W', and T_4 are collinear.

The homothety centered at Q mapping ω_4 to ω_2 must therefore send W' to T_2 (since the tangent to ω_4 at W' is parallel to A_2A_3), so Q, T_4, and T_2 are collinear, as desired.

Clearly, we can translate this into a solution using the Monge-d'Alembert Circle Theorem.

Solution 2. Let A_1A_2 and A_3A_4 meet at Q, A_1A_4 and A_2A_3 at P. We will assume that P, T_1, and T_3 are collinear and prove that so are Q, T_2, and T_4. Consider a circle ω tangent to QT_1 at T_1 and tangent to QT_3. Then T_1 is the insimilicenter of ω_1 and ω and P is the exsimilicenter of ω_1 and ω_3, so according to Monge-d'Alembert's In-Theorem of Problem 79, the insimilicenter of ω and ω_3 lies on PT_1. Furthermore, it lies on the common tangent A_3A_4, but the intersection of these lines is T_3 because of our assumption. Thus T_3 is the insimilicenter of ω and ω_3, and hence ω is tangent to A_3A_4 at T_3. In particular, $QT_1 = QT_3$. Now, calculations involving the Power of a Point Theorem yield $PT_2 = PT_4$. So there is a circle ω' tangent to PT_2 and PT_4 at T_2 and T_4, respectively. Applying the Monge-d'Alembert Theorem as we did before, we get that Q, T_2, and T_4 are collinear.

■

Problem 35. (Cosmin Pohoata, American Mathematical Monthly) Let P be a point inside an arbitrary triangle ABC. Prove that

$$\frac{1}{PA}+\frac{1}{PB}+\frac{1}{PC}\geq\frac{1}{R_a}+\frac{1}{R_b}+\frac{1}{R_c},$$

where R_a, R_b, R_c are the circumradii of triangles PBC, PCA, PAB, respectively.

Solution. We first recall the famous

Erdős-Mordell Inequality. Let P be a point inside triangle ABC and let H_1, H_2, H_3 be the feet of the perpendiculars from P to the sides BC, CA, AB, respectively. Then,

$$PA+PB+PC\geq 2(PH_1+PH_2+PH_3).$$

This was conjectured by Paul Erdős in 1935, and first proved by Mordell in the same year. Several proofs of this inequality have been given, using Ptolemy's Theorem by André Avez, angular computations with similar triangles by Leon Bankoff, area inequality by V. Komornik, or using trigonometry by Mordell and Barrow. Just because they are so beautiful, we will go on a sidenote and present our two favorite proofs.

First Proof. The unusually stringent equality condition should suggest that perhaps the proof proceeds in two stages, with different equality conditions. This is indeed the case.

We will first prove that

$$PA \geq \frac{AB}{BC} \cdot PH_2 + \frac{AC}{BC} \cdot PH_3.$$

As a matter of fact, this step is so important that practically almost all the proofs of the Erdős-Mordell inequality constitute in just different proofs of this key part. So, let's prove it. For example, rewrite it as

$$PA \sin A \geq PH_2 \sin C + PH_3 \sin B,$$

and note that that $PA \sin A = H_2H_3$ (by the Law of Sines in triangle AH_2H_3). On the other hand, the right hand side of the above inequality is the length of the projection of H_2H_3 on BC, and therefore we have equality if and only if H_2H_3 is parallel to the side BC.

Summing up

$$PA \geq \frac{AB}{BC} \cdot PH_2 + \frac{AC}{BC} \cdot PH_3$$

with its other two analogues, yields

$$PA+PB+PC \geq PH_1\left(\frac{CA}{AB}+\frac{AB}{CA}\right)+PH_2\left(\frac{AB}{BC}+\frac{BC}{AB}\right)+PH_3\left(\frac{BC}{CA}+\frac{CA}{AB}\right),$$

with equality holding if and only if the triangles $H_1H_2H_3$ and ABC are homothetic; this holds if and only if P is the circumcenter of triangle ABC. Now for the second step: we note that each of the terms in the parentheses is at least 2 by the AM-GM Inequality. This gives

$$PA + PB + PC \geq 2(PX + PY + PZ),$$

with equality if and only if $AB = BC = CA$, and so our proof is complete.

Now, let's give a proof that doesn't require the often-needed

$$PA \sin A \geq PH_2 \sin C + PH_3 \sin B.$$

Second Proof. We transform it to a trigonometric inequality. Let $h_1 = PH_1$, $h_2 = PH_2$ and $h_3 = PH_3$.

Apply the Law of Sines and the Law of Cosines to obtain

$$\begin{aligned} PA\sin A = H_2H_3 &= \sqrt{{h_2}^2 + {h_3}^2 - 2h_2h_3\cos(180^\circ - A)}, \\ PB\sin B = H_3H_1 &= \sqrt{{h_3}^2 + {h_1}^2 - 2h_3h_1\cos(180^\circ - B)}, \\ PC\sin C = H_1H_2 &= \sqrt{{h_1}^2 + {h_2}^2 - 2h_1h_2\cos(180^\circ - C)}. \end{aligned}$$

So, we need to prove that

$$\sum_{\text{cyclic}} \frac{1}{\sin A}\sqrt{{h_2}^2 + {h_3}^2 - 2h_2h_3\cos(180^\circ - A)} \geq 2(h_1 + h_2 + h_3).$$

The main trouble is that the left hand side has "heavy" terms with square root expressions. Our strategy is to find a lower bound without square roots. To this end, we express the terms inside the square root as the sum of two squares.

$$\begin{aligned} {H_2H_3}^2 &= {h_2}^2 + {h_3}^2 - 2h_2h_3\cos(180^\circ - A) \\ &= {h_2}^2 + {h_3}^2 - 2h_2h_3\cos(B + C) \\ &= {h_2}^2 + {h_3}^2 - 2h_2h_3(\cos B\cos C - \sin B\sin C). \end{aligned}$$

Using $\cos^2 B + \sin^2 B = 1$ and $\cos^2 C + \sin^2 C = 1$, one finds that

$${H_2H_3}^2 = (h_2\sin C + h_3\sin B)^2 + (h_2\cos C - h_3\cos B)^2.$$

Since $(h_2\cos C - h_3\cos B)^2$ is clearly nonnegative, we get

$$H_2H_3 \geq h_2\sin C + h_3\sin B.$$

Hence,

$$\begin{aligned} \sum_{\text{cyclic}} \frac{\sqrt{{h_2}^2 + {h_3}^2 - 2h_2h_3\cos(180^\circ - A)}}{\sin A} &\geq \sum_{\text{cyclic}} \frac{h_2\sin C + h_3\sin B}{\sin A} \\ &= \sum_{\text{cyclic}} \left(\frac{\sin B}{\sin C} + \frac{\sin C}{\sin B}\right) h_1 \\ &\geq \sum_{\text{cyclic}} 2\sqrt{\frac{\sin B}{\sin C}\cdot\frac{\sin C}{\sin B}} h_1 \\ &= 2h_1 + 2h_2 + 2h_3. \end{aligned}$$

This completes the proof.

□

Returning to the problem, let R be a positive number, and let A', B', C' be the images of the points A, B, C under the inversion with respect to the circle with center P and radius R. Then, points A', B', C' lie on the rays PA, PB, PC, respectively, and satisfy $PA' = \frac{R^2}{PA}$, $PB' = \frac{R^2}{PB}$ and $PC' = \frac{R^2}{PC}$.

Now, note that by the Law of Sines, we can rewrite the inequality to prove as

$$\frac{1}{PA} + \frac{1}{PB} + \frac{1}{PC} \geq 2\left(\frac{\sin \angle BPC}{BC} + \frac{\sin \angle CPA}{CA} + \frac{\sin \angle APB}{AB}\right).$$

Since the point P lies inside the triangle $A'B'C'$, we can apply the Erdős-Mordell inequality to the triangle $A'B'C'$ and the point P, and obtain

$$PA' + PB' + PC' \geq 2\,(PX + PY + PZ)\,,$$

where X, Y, Z are the orthogonal projections of the point P on the sides $B'C'$, $C'A'$, $A'B'$ of triangle $A'B'C'$, respectively. Now, in the right-angled triangle PXB', we have

$$PX = PB' \cdot \sin \angle PB'X = PB' \cdot \sin \angle PB'C'.$$

Now, because of $PB' = \frac{R^2}{PB}$ and $PC' = \frac{R^2}{PC}$, we have $\frac{PB'}{PC'} = \frac{PC}{PB}$, which, together with $\angle B'PC' = \angle CPB$, yields that the triangles $B'PC'$ and CPB are similar; therefore, $\angle PB'C' = \angle PCB$. Hence, $PX = PB' \cdot \sin \angle PB'C'$ becomes

$$PX = PB' \cdot \sin \angle PCB = \frac{R^2}{PB} \cdot \sin \angle PCB = R^2 \cdot \frac{\sin \angle PCB}{PB}.$$

By the Law of Sines in triangle BPC, we have $\frac{\sin \angle PCB}{PB} = \frac{\sin \angle BPC}{BC}$, so this becomes $MX = R^2 \cdot \frac{\sin \angle BPC}{BC}$. Similarly, $PY = R^2 \cdot \frac{\sin \angle CPA}{CA}$ and $PZ = R^2 \cdot \frac{\sin \angle APB}{AB}$. Altogether, the inequality

$$PA' + PB' + PC' \geq 2\,(PX + PY + PZ)$$

becomes

$$\frac{R^2}{PA} + \frac{R^2}{PB} + \frac{R^2}{PC} \geq 2\left(R^2 \cdot \frac{\sin \angle BPC}{BC} + R^2 \cdot \frac{\sin \angle CPA}{CA} + R^2 \cdot \frac{\sin \angle APB}{AB}\right),$$

which upon erasing the R^2 becomes precisely what we wanted. This completes the proof.

■

Comment. The inequality above is related to the following problem that appeared once upon a time in an USA Team Selection Test.

Claim. With the same notations, prove that

$$R_a + R_b + R_c \geq PA + PB + PC.$$

The idea for this is, however, much simpler. Just consider the triangle $O_1O_2O_3$ determined by the circumcenters of triangles PBC, PCA, PAB, and apply the Erdős-Mordell inequality for P and this triangle. This yields the claim at once.

Problem 36. (Thebault's Theorem) Through the vertex A of a triangle ABC, a straight line AD is drawn, cutting the side BC at D. Let I be the incenter of triangle ABC, let P be the center of the circle which touches DC, DA, and the circumcircle of ABC internally, and let Q be the center of the circle which touches DB, DA, and the circumcircle of ABC internally. Prove that P, I, Q are collinear.

Solution. The essence of Thebault's result is given by the following Lemma due to Sawayama.

Sawayama's Theorem. Let ABC be a triangle with incenter I and D a point on BC. Consider the circle Γ that is tangent to the circumcircle Ω of ABC internally and also tangent to DC, DA at E, F, respectively. Then, the points E, I, F are collinear.

This idea of proof comes from Jean-Louis Ayme's celebrated paper "Sawayama and Thebault's theorem", where the author gives the first synthetic proof of Thebault's theorem using this Lemma. Here we shall give two completely different proofs for this beautiful result.

First Proof. Denote by Ω the circumcircle of ABC and by Γ the circle tangent to the circumcircle of ABC and lines DC, DA. Let Ω and Γ touch at K. Let M be the midpoint of the arc BC on Ω not conatining K. Then K, E, M are collinear (think: dilation with center K carrying Γ to Ω). Also, A, I, M are collinear, and $MI = MC$.

Let the line EI meet Γ again at F'. It suffices to show that AF' is tangent to Γ. Note that $\angle KF'E$ is subtended by the arc KE in Γ and $\angle KAM$ is subtended by KM in Ω. Since KE and KM are however homothetic with center K, we have $\angle KF'E = \angle KAM$, implying that A, K, I', F' are concyclic.

We have $\angle BCM = \angle CBM = \angle CKM$. So triangles MCE and MKC are similar. Hence $MC^2 = ME \cdot MK$. Since $MC = MI$, we have $MI^2 = ME \cdot MK$, implying that MIE and MKI are in fact similar. Therefore,

$\angle KEI = \angle AIK = \angle AF'K$ (since A, K, I, F' are concyclic). Therefore, AF' is tangent to Ω and the proof is complete.

Second Proof. By Casey's theorem, applied for A, B, C and Γ, we have that

$$AF \cdot BC + AB \cdot CE = AC \cdot BE,$$

or equivalently, $a \cdot AF + ac = (b+c) \cdot BE$, where a, b, c represent the side lengths of triangle ABC.

Let L be the intersection of BC with the internal angle bisector of angle A. We would like to apply Menelaus' theorem in triangle ALD for the three points E, F, I lying on its sidelines; thus, we want to show that

$$\frac{EL}{EF} \cdot \frac{FD}{FA} \cdot \frac{IA}{IL} = 1,$$

which is the same thing as proving that

$$\frac{EL}{FA} \cdot \frac{IA}{IL} = 1$$

because $DE = DF$. However, using Van Aubel's theorem, we know that $\frac{IA}{IL} = \frac{b+c}{a}$, thus, it suffices to show that $a \cdot AF = LE \cdot (b+c)$. However, by the Angle Bisector theorem, we know that $BL = \frac{ac}{b+c}$, thus, since

$$LE = BE - BL = BE - \frac{ac}{b+c},$$

we can conclude that the relation we got from Casey, i.e. $a \cdot AF + ac = (b+c) \cdot BE$, implies $a \cdot AF = LE \cdot (b+c)$. This completes the proof.

□

Returning to our proof of Thebault's theorem, we notice that Sawayama's theorem yields the perfect setting for a Pappus' theorem, à la Solution 8 of Problem 14.

Let (P), (Q) be the two Thebault circles. Suppose (P) meets DC at E and DA at H and suppose that (Q) meets DB at G and DA at F. By the Lemma, we have that E, I, H are collinear and that G, F, I are collinear. Moreover, note that $DP \perp EH$, $DQ \perp GF$ and $DP \perp DQ$. Hence, it follows that $DP \| GF$ and $DQ \| EH$. Now, working in the projective plane and thinking of parallel lines as just lines meeting on the line at infinity, let I_0, I_1, I_2 be the infinity points $I_0 = GQ \cap EP$, $I_1 = DP \cap GF$, $I_2 = DQ \cap EH$. By Pappus' theorem for the triplets G, E, D and I_2, I_1, I_0, the intersections

$$GI_1 \cap EI_2 = GI \cap EH = I,$$

$$EI_0 \cap DI_1 = EP \cap DP = P,$$

$$GI_0 \cap DI_2 = GQ \cap DQ = Q$$

are collinear. This proves Thebault's Theorem.

■

Comment 1. Note that both Sawayama's theorem and Thebault's theorem admit versions for the case when the internal tangencies are replaced by external tangencies. We record these below for reference purposes.

External Sawayama's Theorem. Let ABC be a triangle with $A-$excenter I_a and D a point on BC. Consider the circle Γ that is tangent to the circumcircle Ω of ABC externally and also tangent to the lines DC, DA at E, F, respectively. Then, the points E, I_a, F are collinear.

External Thebault's Theorem. Through the vertex A of a triangle ABC, a straight line AD is drawn, cutting the side BC at D. Let I_a be the A-excenter of triangle ABC, let P be the center of the circle which touches DC, DA, and the circumcircle of ABC externally, and let Q be the center of the circle which touches DB, DA, and the circumcircle of ABC externally. Prove that P, I_a, Q are collinear.

Comment 2. Another interesting property which we shall use later is the following beautiful identity.

Theorem. Maintaining the notations from Problem 36, and furthermore denoting $\angle ADB$ by ϕ, the inradius of ABC by r, the radius of (P) by r_1, and the radius of (Q) by r_2, we have that

$$r = r_1 \cos^2 \frac{\phi}{2} + r_2 \sin^2 \frac{\phi}{2}.$$

This follows immediately from adding the Casey relation we obtained in the second proof above with the relation we get from applying Casey to the other circle. If we denote as above by E and G the tangency points of BC with (P) and (Q), respectively, this addition yields

$$DE = \frac{a}{s} \cdot AD,$$

where s is the semi-perimeter of triangle ABC. The rest is just trigonometry, which we leave for the reader.

Problem 37. (Jean-Pierre Ehrmann and Cosmin Pohoata, MathLinks Contest 2007) Let ABC be a triangle with circumcircle Γ and let A' be a point on the side BC. Let $\mathcal{T}_1$ and $\mathcal{T}_2$ be the circles tangent simultaneously to AA', BA', Γ and AA', CA', Γ, respectively. Prove that

(a) If A' is the foot of the internal angle bisector from A, then $\mathcal{T}_1$ and $\mathcal{T}_2$ are tangent to each other at the incenter of triangle ABC.

(b) If A' is the tangency point of the A-excircle with BC, then $\mathcal{T}_1$ and $\mathcal{T}_2$ are congruent.

Solution. (a) Note that Sawayama's theorem from the proof of Problem 36 trivializes the problem. Indeed, let the circles $\mathcal{T}_1$, $\mathcal{T}_2$, centered at I_1 and I_2, respectively, touch the side BC at D and E, and the segment AA' at R and S, respectively. From Sawayama's Theorem, the incenter I of triangle ABC lies on the lines RD and SE. However, I also lies on the cevian AA'. Thus $R \equiv S \equiv I$, and so circles $\mathcal{T}_1$ and $\mathcal{T}_2$ are both tangent to AA' at I, which proves the claim.

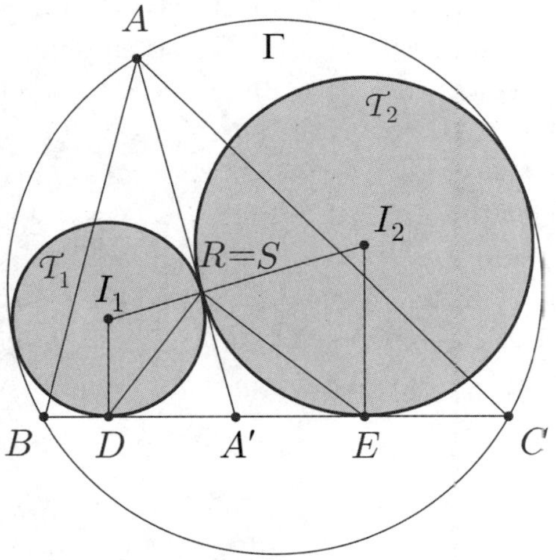

We shall now give a more fun proof using a very nice Lemma.

Lemma. Let ABC be an isosceles triangle with $AB = AC$. Let Γ be its circumcircle and let ω be an arbitrary circle, which is tangent to the side BC and to the arc BC that not contain the vertex A. Then, the power of A with respect to ω is constant, independent of the choice of ω.

Proof of Lemma. Note that the power of A with respect to ω is just the squared length t of the tangent from A to ω. Hence, it suffices to compute t and prove it is constant. Let X be the tangency point of ω with BC.

By Casey's theorem, applied for "quadrilateral" $AB\omega C$, we have that

$$\begin{aligned} t \cdot BC &= AB \cdot CX + AC \cdot BX \\ &= AB \cdot (CX + BX) \\ &= AB \cdot BC. \end{aligned}$$

Hence $t = AB$, which is independent of the choice of ω. This proves the Lemma.

Returning to the problem, let M be the second intersection of the bisector AA' with the circumcircle Γ. This point is the midpoint of the arc BC, so $MB = MC$. In particular, triangle MBC is isosceles and circles $\mathcal{T}_1$ and $\mathcal{T}_2$ are circles tangent to BC and the circumcircle Γ of triangle MBC. Hence, from the above Lemma, the tangents MR and MS satisfy $MR = MS = MB = MC$. However, it is well known that $MB = MC = MI$. This proves that $MR = MS = MI$, and so it follows that $R \equiv S \equiv I$, as desired.

(b) As before, let the circles $\mathcal{T}_1(I_1)$, $\mathcal{T}_2(I_2)$ touch the side BC at D and E, and the segment AA' at R and S, respectively. From Sawayama's theorem, the incenter I lies on the lines RD and SE, so the lines ID and IE are perpendicular to I_1A' and I_2A', respectively. Moreover, the internal and external angle bisector of angle $AA'B$, so $I_1A' \perp I_2A'$, as well.

From Thebault's theorem, I_1I_2 passes through I; hence, if circles $\mathcal{T}_1$ and $\mathcal{T}_2$ are congruent, then the line containing I_1, I_2, I is parallel to BC, which means that quadrilateral I_1DEI_2 is a rectangle. Thus, if X is the tangency point of the incircle (I) of triangle ABC with BC, then II_1DX and II_2EX are rectangles, which yields $ID = XI_1$ and $IE = XI_2$. In particular, this implies that triangles IDE and XI_1I_2 are congruent. Consequently,

$$\angle I_1XI_2 = \angle DIE = 90^\circ.$$

This proves that $\angle I_1XI_2 = \angle I_1A'I_2$, therefore $I_1I_2A'X$ is cyclic. However, as argued before, I_1I_2 is parallel to $BC \equiv A'X$. Hence, $I_1I_2A'X$ is actually an isosceles trapezoid with $I_1X = I_2A'$. It follows that triangles I_1DX and I_2EA' are congruent, which makes $DX = EA'$. But this yields $BX = CA'$, and so we get that A' is the reflection of X across the midpoint of side BC. It is well known that this implies A' is the tangency point of the A-excircle with BC. For the converse, note that we are able to reverse all the above steps to get that the line, which passes through the centers of the Thebault circles, is parallel to BC. This completes the proof.

■

Comment. The first part of the problem was given before in an old Romanian IMO team selection test.

Problem 38. (Cosmin Pohoata and Josef Tkadlec, Mathematical Reflections) Suppose M and N are distinct points in the plane of triangle ABC such that $AM : BM : CM = AN : BN : CN$. Prove that the line MN contains the circumcenter of triangle ABC.

Solution 1. Let the circumcenter of the circle be O. Now, consider the inversion Ψ with center O and power R^2, where R is the circumradius of triangle ABC. Let this inversion map M to M' and N to N'. We will first show that N' can be M. The inversion gives that triangles ONB and OBN' are similar, which implies that $\frac{BN}{BN'} = \frac{ON}{R}$. Similarly,

$$\frac{AN}{AN'} = \frac{ON}{R} = \frac{CN}{CN'} = \frac{BN}{BN'}.$$

This means that M satisfies the same conditions as N', so N' could equal M.

Now, we will show that for each N, there is only one M, which means that N' would have to equal M. Invert around N with arbitrary power. Let this map A to A', B to B', M to M_1, and C to C'. The inversion gives that triangles $NB'M_1$ and NMB are similar, so $\frac{NB}{MB} = \frac{M_1N}{M_1B'}$, and similarly, $\frac{NC}{MC} = \frac{M_1N}{M_1C'}$. This means that

$$\frac{M_1N}{M_1C'} = \frac{NC}{MC} = \frac{NB}{MB} = \frac{M_1N}{M_1B'};$$

hence $M_1B' = M_1C'$. Therefore, M_1 lies on the perpendicular bisector of $B'C'$. Similarly, M_1 lies on the perpendicular bisector of $A'B'$, so M_1 is the circumcenter of triangle ABC. This means that only one possible M_1 exists, and therefore, only one possible M exists. Hence, we see that $N' = M$. Then, the inversion with center O and radius OB maps N to M, so it follows that M and N lie on a line passing through O. This completes the proof

While Solution 1 presents the way the result was actually discovered, Solution 2 is the "proof from the book" that will make you remember it.

Solution 2. Recall that given a positive real number r, the Apollonius circle (XY, r) of segment XY is the circle centered on XY that represents the locus of the points P in plane such that $\frac{PX}{PY} = r$. Consequently, if we rewrite the proportionality from the hypothesis as

$$\frac{AM}{AN} = \frac{BM}{BN} = \frac{CM}{CN} := r > 0,$$

we get that the points A, B, C all lie on the Appollonius circle (MN, r) of segment MN. The circumcenter O of triangle ABC is the center of this circle, and so it lies on line MN.

■

Problem 39. (Romania IMO TST 2007) The points M, N, P are chosen on the sides BC, CA, AB of a triangle ABC, such that the triangle MNP is acute-angled. We denote with x the length of the shortest altitude of the triangle ABC, and with y the length of the longest altitude of the triangle MNP. Prove that $x \leq 2y$.

Solution. Let H be the orthocenter of triangle MNP and A', B', C' the feet of the perpendiculars from H to BC, CA, AB, respectively. We have that

$$\frac{1}{x} \cdot \sum_{\text{cyclic}} HA' \geq \sum_{\text{cyclic}} \frac{HA'}{h_a} = \sum_{\text{cyclic}} \frac{[HBC]}{[ABC]} = 1.$$

Therefore

$$x \leq HA' + HB' + HC'.$$

Furthermore, note that $\sum_{\text{cyclic}} HA' \leq \sum_{\text{cyclic}} HM$. Hence, it suffices to prove that

$$\sum_{\text{cyclic}} HM \leq 2y.$$

Suppose that MM_1 is the longest altitude of $\triangle MNP$; therefore NP is the shortest side. Call D the reflection of H across M_1. We know that D is on the circumcircle of triangle MNP. Furthermore, observe that $\sum_{\text{cyclic}} HM = HM + DN + DP$, hence $\sum_{\text{cyclic}} HM \leq 2y$ rewrites as $DN + DP \leq MD$.

On the other hand, Ptolemy's theorem in the cyclic quadrilateral $MNDP$ gives

$$\begin{aligned} MD \cdot NP &= DN \cdot MP + DP \cdot MN \\ &\geq DN \cdot NP + DP \cdot NP. \end{aligned}$$

Hence $MD \geq DN + DP$, as claimed. This complete the proof.

■

Problem 40. (Lev Emelyanov, Forum Geometricorum and Romania IMO TST 2009) Let ABC be a nonisosceles triangle, for which denote by X, Y, Z the tangency points of its incircle with the sides BC, CA and AB, respectively. Let D be the intersection of OI with the sideline BC, where O, I are the circumcenter and incenter, respectively. The line perpendicular to YZ through X cuts AD at E. Prove that the line YZ is the perpendicular bisector of EX.

Solution. We recall the following known fact about the orthocenter of triangle XYZ that surfaced in a lot of contest problems.

Lemma. The orthocenter of XYZ lies on the line OI.

As you might realize after working on this for a couple of minutes, the result is not that easy to approach. We give two completely different proofs.

First Proof of Lemma. Let A', B', C' be the midpoints of YZ, ZX and XY, respectively. Note that

$$IA \cdot IA' = \left(\frac{r}{\sin\frac{A}{2}}\right) \cdot \left(r \sin\frac{A}{2}\right) = r^2.$$

Similarly $IB \cdot IB' = IC \cdot IC' = r^2$. Therefore, the inversion $\Psi(I, r^2)$ with respect to the incircle of ABC maps the vertices A, B, C to A', B', C', and moreover maps the circumcircle of ABC into the circumcircle of $A'B'C'$. But if a circle with center P is inverted in a circle with center Q, then its image is a circle whose center lies on the line PQ. Thus, the circumcenter O' of $A'B'C'$ lies on the line OI. Since $A'B'C'$ is the medial triangle of XYZ, its circumcenter O' is the nine-point center of XYZ. Following the fact that I is the circumcenter of XYZ, it becomes clear now that OI is the Euler line of XYZ. Obviously, the orthocenter H' of XYZ lies on this line.

□

The second proof uses a simpler homothety argument.

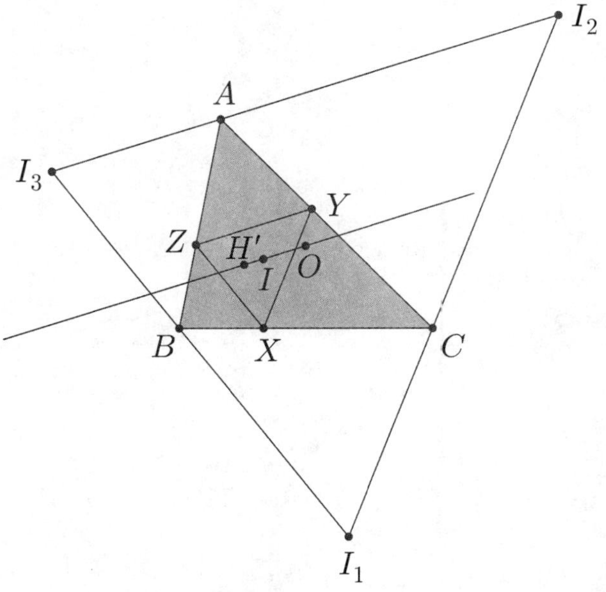

Second Proof of Lemma. Let $I_aI_bI_c$ be the excentral triangle of ABC. The lines YZ and I_bI_c are parallel because both are perpendicular to AI. Similarly, $ZX \| I_cI_a$ and $XY \| I_aI_b$. Hence, the excentral triangle and the intouch triangle are homothetic and their Euler lines are parallel. However, I and O are the orthocenter and the nine-point center of the excentral triangle, whereas I is the circumcenter of the intouch triangle. Therefore, the line OI is their common Euler line, and thus it contains the orthocenter H' of XYZ. This completes the proof.

□

Returning to the problem, we may now rewrite it in the following equivalent version.

Claim. Let E be the reflection of X into the line YZ. Denote by D the intersection of AE with the sideline BC. Prove that D lies on the line OI.

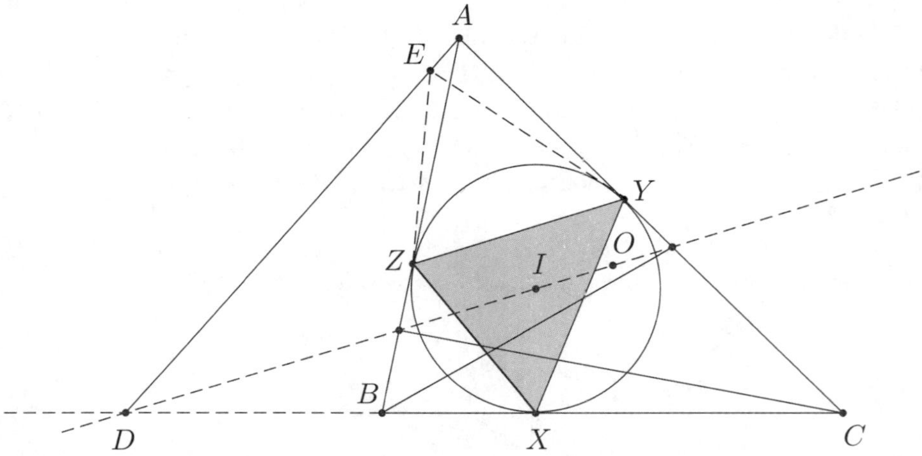

Proof of Claim. Denote by T the intersection of AI with BC. Since XE and AT are both perpendicular to YZ, they are parallel and therefore the collinearity is equivalent to $EH'/H'X = AI/IT$.

It is known that

$$\frac{AI}{IT} = \frac{CA + AB}{BC} = \frac{\sin B + \sin C}{\sin A}.$$

For example, this follows from Van Aubel's theorem.

For any acute triangle, $AH = 2R \cos A$. The angles of the intouch triangle are

$$X = \frac{B + C}{2}, \quad Y = \frac{C + A}{2}, \quad Z = \frac{A + B}{2}.$$

It is clear that XYZ is always acute, and

$$XH' = 2r \cos X = 2r \cos \frac{B + C}{2} = 2r \sin \frac{A}{2}.$$

Thus,

$$\begin{aligned}
\frac{EH'}{H'X} &= \frac{EX - H'X}{H'X} = \frac{EX \cdot YZ}{H'X \cdot YZ} - 1 \\
&= \frac{4[XYZ]}{H'X \cdot YZ} - 1 \\
&= \frac{2r^2(\sin 2X + \sin 2Y + \sin 2Z)}{2r \sin X \cdot 2r \cos X} - 1 \\
&= \frac{\sin 2Y + \sin 2Z}{\sin 2Y} \\
&= \frac{\sin B + \sin C}{\sin A}.
\end{aligned}$$

This proves the claim and the problem. Note also that the above works shows that the line joining D to the projection of X on YZ passes through the midpoint of the bisector of angle A.

Moreover, note that the line OI is parallel to BC if and only if the projection of X on YZ lies on the line joining the midpoints of AB and AC. ■

Problem 41. (Bogdan Enescu, IMO 2005) Six points are chosen on the sides of an equilateral triangle ABC: A_1, A_2 on BC, B_1, B_2 on CA and C_1, C_2 on AB, such that they are the vertices of a convex hexagon $A_1A_2B_1B_2C_1C_2$ with equal side lengths. Prove that the lines A_1B_2, B_1C_2 and C_1A_2 are concurrent.

Solution. The idea is to use the following very simple result about similar triangles and vectors.

Claim. Let ABC be a triangle and let u, v, w be vectors that sum up to 0 (and thus form a triangle). Then either of the following statements implies that the triangle formed by u, v, w is similar to triangle ABC:

(a) $\frac{|u|}{AB} = \frac{|v|}{BC} = \frac{|w|}{CA}$;

(b) $u \| \overrightarrow{AB}$, $v \| \overrightarrow{BC}$, $w \| \overrightarrow{CA}$;

Conversely, a) and b) imply that the vectors u, v, w form a triangle.

Indeed, both statements follow from the similarity criteria for two triangles. Note that (a) follows from the side similarity criterion and (b) from the angle similarity criterion (i.e., two triangles with parallel sides are similar). The converse follows from the construction of triangle T formed by the vectors u, v and $w' = -u - v$. We have $\frac{|u|}{AB} = \frac{|v|}{BC}$ and the angle between u and v equals the angle between $\overrightarrow{AB}$ and $\overrightarrow{AC}$. Hence T is similar to ABC which implies

$$\frac{|w'|}{CA} = \frac{|u|}{AB} = \frac{|w|}{CA} \text{ and } w \| \overrightarrow{CA} \| w',$$

so w and w' have the same length and the same direction, thus $w = w' = -u - v$, implying that $u + v + w = 0$, or in other words, u, v, w form a triangle, as desired. This proves the Claim.

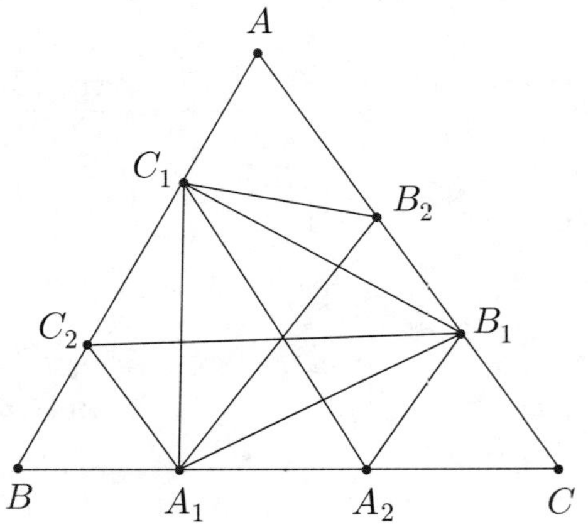

Returning to the problem, we notice that the vectors $\overrightarrow{A_1A_2}$, $\overrightarrow{B_1B_2}$, $\overrightarrow{C_1C_2}$ are equal in length and parallel to the sides of triangle ABC (which are also equal in length); thus, by the converse of the Claim above, we can deduce that their sum is 0. Also, note that

$$\begin{aligned} 0 &= \overrightarrow{A_1A_2} + \overrightarrow{A_2B_1} + \overrightarrow{B_1B_2} + \overrightarrow{B_2C_1} + \overrightarrow{C_1C_2} + \overrightarrow{C_2A_1} \\ &= \overrightarrow{A_2B_1} + \overrightarrow{B_2C_1} + \overrightarrow{C_2A_1}. \end{aligned}$$

But the vectors $\overrightarrow{A_2B_1}$, $\overrightarrow{B_2C_1}$, $\overrightarrow{C_2A_1}$ also have the same length. Thus, we can apply the Claim again to see that they form an equilateral triangle, and so they are rotations of each other (by 120° and 240°), as we know $\overrightarrow{A_1A_2}$, $\overrightarrow{B_1B_2}$, $\overrightarrow{C_1C_2}$ are. Hence triangles $A_1A_2B_1$, $B_1B_2C_1$, $C_1C_2A_1$ are also rotations of each other so they are congruent. Thus, $A_1B_1 = B_1C_1 = C_1A_1$ and so $A_1B_1C_1$ is equilateral. As $C_1B_2 = B_2B_1$, we deduce that A_1, B_2 lie on the perpendicular bisector of B_1C_1, so A_1B_2 is the perpendicular bisector of B_1C_1. Analogously, B_1C_2 and C_1B_2 are also perpendicular bisectors in triangle $A_1B_1C_1$, so the three lines A_1B_2, B_1C_2 and C_1A_2 are concurrent. This completes the proof. ■

Problem 42. (Victor Oxman, Crux Mathematicorum) Given triangle ABC, its centroid G, and its incenter I, construct, using only an unmarked straightedge, its orthocenter H.

Solution. We will first establish a Lemma.

Lemma. Given a line segment and its midpoint, and any other point O in the plane, a line passing through O and parallel to the given line segment may be constructed using only an unmarked straightedge.

Proof. Let AD be the given line segment and let K be its midpoint. Let P be any point on AO extended. Connect P with D and with K. Draw segment OD. Let M be the intersection of OD and PK, and let N be the intersection of AM with PD. We will show that ON is parallel to AD.

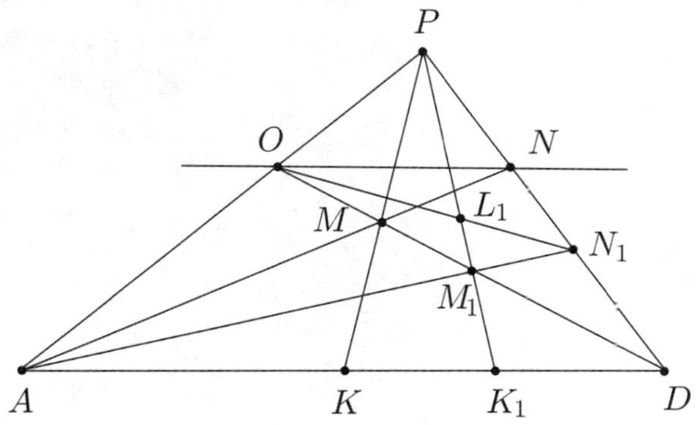

Suppose instead that ON_1 is parallel to AD, with N_1 on PD. Let M_1 be the intersection of AN_1 and OD, let K_1 be the intersection of PM_1 and AD, and let L_1 be the intersection of ON_1 with PK_1. Then, $\triangle AM_1K_1 \sim \triangle N_1M_1L_1$ and $\triangle OM_1L_1 \sim \triangle DM_1K_1$. Therefore,

$$\frac{AK_1}{L_1N_1} = \frac{K_1M_1}{M_1L_1} = \frac{K_1D}{OL_1}.$$

Also $\triangle APK_1 \sim \triangle OPL_1$ and $\triangle DK_1P \sim \triangle N_1L_1P$, which implies

$$\frac{AK_1}{OL_1} = \frac{K_1P}{L_1P} = \frac{K_1D}{L_1N_1}.$$

Combining these two relations, we get that

$$AK_1 \cdot OL_1 = L_1N_1 \cdot K_1D \quad \text{and} \quad AK_1 \cdot L_1N_1 = OL_1 \cdot K_1D.$$

Consequently,

$$(AK_1)^2 \cdot OL_1 \cdot L_1N_1 = (K_1D)^2 \cdot OL_1 \cdot L_1N_1,$$

from which we have $(AK_1)^2 = (K_1D)^2$, and $AK_1 = K_1D$. Thus point K_1 coincides with point K, PK_1 coincides with PK, point M_1 with M and point N_1 with N. Thus ON is parallel to AD. This proves the Lemma.

□

Using the centroid G, draw the medians AA_1, BB_1, CC_1, where A_1, B_1, C_1 are the midpoints of BC, CA, AB, respectively. Similarly, using the incenter I, draw the angle bisectors AA_1, BB_2, CC_2, where A_2, B_2, C_2 lie on BC, CA, AB, respectively. Draw triangle $A_1B_1C_1$ and denote $C_1A_1 \cap BB_2$ by B_3, $A_1B_1 \cap CC_2$ by C_3, $B_1C_1 \cap AA_2$ by A_3. Thus, A_3, B_3, C_3 are the midpoints of AA_1, BB_2, CC_2, respectively.

By applying the Lemma twice, we may draw a straight line BE through B parallel to angle bisector CC_2 and draw a straight line BF parallel to angle bisector AA_2 (with points E and F located on the line CA). We see that triangles FAB and BCE are isosceles ($\angle FBA = \angle BFA$, $\angle EBC = \angle BEC$). Draw a straight line A_1K parallel to CA with K on BE. Thus K is the midpoint of BE. Similarly, get the midpoint N of FB. Thus $AN \perp FB$ and $CK \perp BE$.

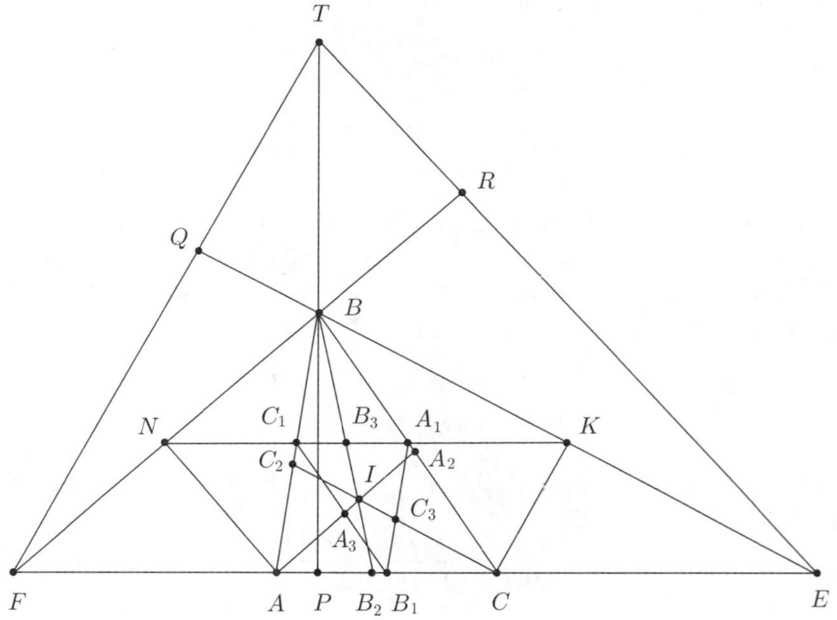

Now, we may draw lines parallel to FB and BE through the points C_1 and A_1, respectively, and get the midpoints of AN and CK. Then, applying the Lemma, we may draw FT parallel to CK and ET parallel to AN. Denote $ET \cap FB$ by R and $BE \cap FT$ by Q. Thus, $FR \perp ET$, $RQ \perp FT$, and point B is the point of intersection of the altitudes of triangle FTE. Draw line TB, denote $TB \cap CA$ by P and get $BP \cap CA$; that is, BP is an altitude of triangle ABC. We can construct a second altitude similarly. Their intersection is H, the orthocenter. This completes the proof.

■

Problem 43. (Romania IMO TST 2009) Let ABC be a scalene triangle and let Ω be a circle that intersects the side BC at A_1, A_2, the side CA at B_1, B_2, and the side AB at C_1, C_2. Let the tangents at A_1 and A_2 with respect to Ω meet at P, and similarly define R and S. Prove that lines AP, BQ, CR are concurrent.

Solution 1. By Salmon's theorem, we can write

$$\frac{RO}{PO} = \frac{\delta(R,a)}{\delta(P,c)} \quad \text{and} \quad \frac{QO}{PO} = \frac{\delta(Q,a)}{\delta(P,b)}.$$

Therefore,

$$\frac{\delta(P,c)}{\delta(P,b)} = \frac{\delta(R,a)}{\delta(Q,a)} \cdot \frac{QO}{RO}.$$

Analogously, we get that

$$\frac{\delta(Q,a)}{\delta(Q,c)} = \frac{\delta(P,b)}{\delta(R,b)} \cdot \frac{RO}{PO} \quad \text{and} \quad \frac{\delta(R,b)}{\delta(R,a)} = \frac{\delta(Q,c)}{\delta(P,c)} \cdot \frac{PO}{QO}.$$

Multiplying these together yields

$$\begin{aligned} \frac{\delta(P,c)}{\delta(P,b)} \cdot \frac{\delta(Q,a)}{\delta(Q,c)} \cdot \frac{\delta(R,b)}{\delta(R,a)} &= \frac{\delta(R,a)}{\delta(Q,a)} \cdot \frac{\delta(P,b)}{\delta(R,b)} \cdot \frac{\delta(Q,c)}{\delta(P,c)} \cdot \frac{QO}{RO} \cdot \frac{RO}{PO} \cdot \frac{PO}{QO} \\ &= \left(\frac{\delta(P,c)}{\delta(P,b)} \cdot \frac{\delta(Q,a)}{\delta(Q,c)} \cdot \frac{\delta(R,b)}{\delta(R,a)} \right)^{-1}. \end{aligned}$$

Thus,

$$\frac{\delta(P,c)}{\delta(P,b)} \cdot \frac{\delta(Q,a)}{\delta(Q,c)} \cdot \frac{\delta(R,b)}{\delta(R,a)} = 1,$$

and so it follows that lines AP, BQ, CR are concurrent, by Trigonometric Ceva.

■

Solution 2. The second proof uses Sondat's theorem from Problem 100, which we have not solved yet, since it is a result more difficult than the problem at hand. Nonetheless, let us see how we could apply it. We consider poles and polars with respect to circle $\Omega(O)$. Note that BC is the polar of P, CA is the polar of Q, and AB is the polar of R. In particular, this means that $OP \perp BC$, $OQ \perp CA$, and $OR \perp AB$, respectively.

On the other hand, $A = CA \cap AB$ is the pole of QR, $B = AB \cap BC$ is the pole of RP, and $C = BC \cap CA$ is the pole of PQ. Hence, $OA \perp QR$, $OB \perp RP$, and $OC \perp PQ$, respectively. In other words, this means that triangles ABC and PQR are orthologic with one common orthology center. By Problem 100, part (a), we conclude that lines AP, BQ, CR are concurrent.

■

Problem 44. (IMO Shortlist 2009) Let the sides AD and BC of the quadrilateral $ABCD$ (such that AB is not parallel to CD) intersect at point P. Points O_1 and O_2 are circumcenters and points H_1 and H_2 are orthocenters of triangles ABP and CDP, respectively. Denote the midpoints of segments O_1H_1 and O_2H_2 by E_1 and E_2, respectively. Prove that the perpendicular from E_1 on CD, the perpendicular from E_2 on AB and the lines H_1H_2 are concurrent.

Solution 1. We keep triangle ABP fixed and move the line CD parallel to itself uniformly, i.e., linearly dependent on a single parameter. Then the points C and D also move uniformly. Hence, the points O_2, H_2 and E_2 move uniformly, too. Therefore also the perpendicular from E_2 on AB moves uniformly. Obviously, the points O_1, H_1, E_1 and the perpendicular from E_1 on CD do not move at all. Hence, the intersection point S of these two perpendiculars moves uniformly. Since H_1 does not move, while H_2 and S move uniformly along parallel lines (both are perpendicular to CD), it suffices to prove their collinearity for two different positions of CD.

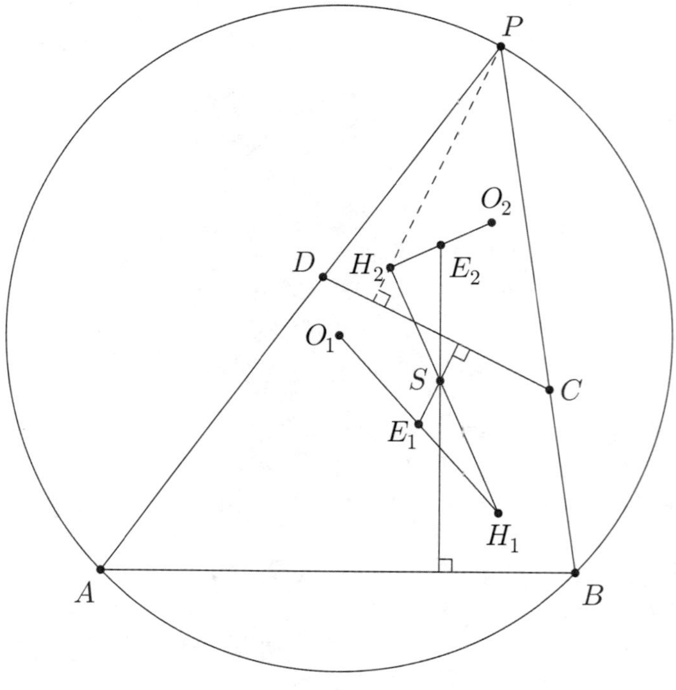

Let CD pass through either point A or point B. Note that by hypothesis these two cases are different. We will consider the case $A \in CD$, i.e., $A = D$. So we have to show that the perpendiculars from E_1 on AC and from E_2 on AB intersect on the altitude AH of triangle ABC.

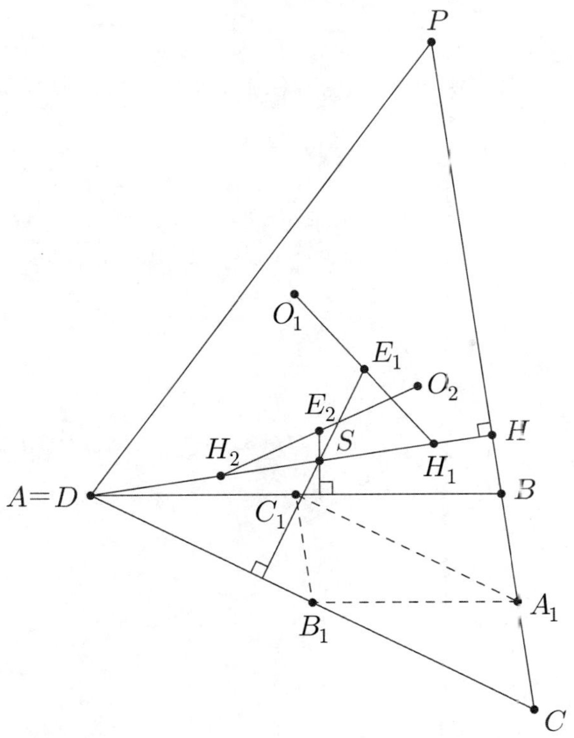

To this end, we consider the midpoints A_1, B_1, C_1 of BC, CA, AB, respectively. As E_1 is the nine-point center of ABP, we have $E_1C_1 = E_1H$. Similarly, $E_2B_1 = E_2H$. Note further that a point X lies on the perpendicular from E_1 on A_1C_1 if and only if

$$XC_1^2 - XA_1^2 = E_1C_1^2 - E_1A_1^2.$$

Similarly, the perpendicular from E_2 on A_1B_1 is characterized by

$$XA_1^2 - XB_1^2 = E_2A_1^2 - E_2B_1^2.$$

The line H_1H_2, which is perpendicular to B_1C_1 and contains A, is given by

$$XB_1^2 - XC_1^2 = AB_1^2 - AC_1^2.$$

Now, the three lines are concurrent if and only if

$$\begin{aligned} 0 &= XC_1^2 - XA_1^2 + XA_1^2 - XB_1^2 + XB_1^2 - XC_1^2 \\ &= E_1C_1^2 - E_1A_1^2 + E_2A_1^2 - E_2B_1^2 + AB_1^2 - AC_1^2 \\ &= -E_1A_1^2 + E_2A_1^2 + E_1H^2 - E_2H^2 + AB_1^2 - AC_1^2, \end{aligned}$$

i.e., it suffices to show that

$$E_1A_1^2 - E_2A_1^2 - E_1H^2 + E_2H^2 = \frac{AC^2 - AB^2}{4}.$$

We have

$$\frac{AC^2 - AB^2}{4} = \frac{HC^2 - HB^2}{4} = \frac{(HC + HB)(HC - HB)}{4} = \frac{HA_1 \cdot BC}{2},$$

where the first equality holds because $AH \perp BC$.

Let F_1, F_2 be the projections of E_1, E_2 on BC. Obviously, these are the midpoints of HP_1, HP_2, where P_1 ,P_2 are the midpoints of PB and PC, respectively. Then

$$\begin{aligned}
& E_1A_1^2 - E_2A_1^2 - E_1H^2 + E_2H^2 \\
= \ & F_1A_1^2 - F_1H^2 - F_2A_1^2 + F_2H^2 \\
= \ & (F_1A_1 - F_1H)(F_1A_1 + F_1H) - (F_2A_1 - F_2H)(F_2A_1 + F_2H) \\
= \ & A_1H \cdot (A_1P_1 - A_1P_2) \\
= \ & \frac{A_1H \cdot BC}{2} \\
= \ & \frac{AC^2 - AB^2}{4},
\end{aligned}$$

which completes the proof.

Solution 2. Let the perpendicular from E_1 on CD meet PH_1 at X, and the perpendicular from E_2 on AB meet PH_2 at Y (see below). Let ϕ be the intersection angle of AB and CD. Denote by M, N the midpoints of PH_1, PH_2, respectively.

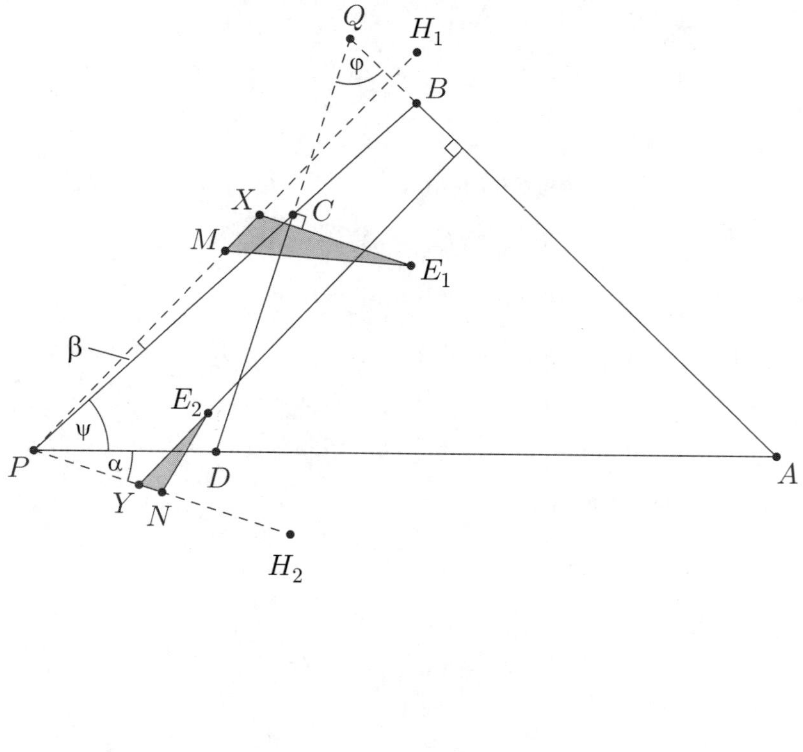

We will prove now that triangles E_1XM and E_2YN have equal angles at E_1, E_2, and supplementary angles at X, Y. In the following, angles are understood as oriented, and equalities of angles modulo 180°. Let $\alpha = \angle H_2PD$, $\Phi = \angle DPC$, $\beta = \angle CPH_1$. Then $\alpha + \Phi + \beta = \phi$, $\angle E_1XH_1 = \angle H_2YE_2 = \phi$, thus $\angle MXE_1 + \angle NYE_2 = 180°$.

By considering the Feuerbach circle of triangle ABP whose center is E and that goes through M, we have $\angle E_1MH_1 = \Phi + 2\beta$. Analogous considerations with the Feuerbach circle of triangle DCP yield $\angle H_2NE_2 = \Phi + 2\alpha$. Hence indeed

$$\angle XE_1M = \phi - (\Phi + 2\beta) = (\Phi + 2\alpha) - \phi = \angle YE_2N.$$

It follows now that

$$\frac{XM}{ME_1} = \frac{YN}{NE_2}.$$

Furthermore, ME_1 is half the circumradius of triangle ABP, while PH_1 is the distance of P to the orthocenter of that triangle, which is twice the circumradius times the cosine of Φ. Together with analogous reasoning for triangle DCP, we have

$$\frac{ME_1}{PH_1} = \frac{1}{4\cos\Phi} = \frac{NE_2}{PH_2}.$$

By multiplication, we get

$$\frac{XM}{PH_1} = \frac{YN}{PH_2}, \quad \text{and so} \quad \frac{PX}{XH_1} = \frac{H_2Y}{YP}.$$

Let E_1X, E_2Y meet H_1H_2 in R, S, respectively. Since

$$\frac{H_2R}{RH_1} = \frac{PX}{XH_1} \quad \text{and} \quad \frac{H_2S}{SH_1} = \frac{H_2Y}{YP},$$

it follows that R and S coincide. This completes the proof.

■

Problem 45. (Cosmin Pohoata, Mathematical Reflections) Let ABC be an acute-angled triangle with circumcircle $\Gamma(O)$ and let ℓ be a line in the plane which intersects the lines BC, CA, AB at X, Y, Z, respectively. Let ℓ_A, ℓ_B, ℓ_C be the reflections of ℓ across BC, CA, AB, respectively. Furthermore, let M be the Miquel point of triangle ABC with respect to line ℓ.

(a) Prove that lines ℓ_A, ℓ_B, ℓ_C determine a triangle whose incenter lies on the circumcircle of triangle ABC.

(b) If S is the incenter from (a) and O_a, O_b, O_c denote the circumcenters of triangles AYZ, BZX, CXY, respectively, prove that the circumcircles of triangles SOO_a, SOO_b, SOO_c are concurrent at a second point, which lies on Γ.

Solution. (a) Let $A' = \ell_B \cap \ell_C$, $B' = \ell_C \cap \ell_A$ and $C' = \ell_A \cap \ell_B$. Note first that Y is the point where ℓ, ℓ_B, AC concur, or since ℓ, ℓ_B are symmetric around AC, they are also symmetric around the perpendicular to AC at Y, and we have $\angle ZYA' = 180° - 2\angle AYZ$, and similarly $\angle YZA' = 180° - 2\angle AZY$, hence

$$\angle YA'Z = 2(\angle AYZ + \angle AZY) - 180° = 180° - 2\angle YAZ = 180° - \angle YO_aZ,$$

hence YO_aZA' is cyclic. Moreover,

$$\angle YO_aA' = \angle YZA' = 180° - 2\angle AZY = 180° - \angle AO_aY,$$

hence A, O_a, A' are collinear. Finally, $\angle YA'A = \angle YA'O_a = \angle YZO_a$, and similarly $\angle ZA'A = \angle ZYO_a$, hence AA' is the angle bisector of $\angle YA'Z$, hence of $\angle B'A'C'$. Or the incenter of $A'B'C'$ is the point where AO_a, BO_b, CO_c concur.

Let S_a, S_b be the points where AO_a, BO_b meet the circumcircle of ABC for the second time. Clearly,

$$\angle CAS_a = \angle YAO_a = 90° - \frac{1}{2}\angle AO_aY = 90° - \angle AZY,$$

and similarly $\angle CBS_b = 90° - \angle BZX$. But $\angle AZY = \angle BZX$ since it is the angle between ℓ and AB, or $S_a = S_b$. Similarly, $S_b = S_c$, or AO_a, BO_b, CO_c meet at a point $S = S_a = S_b = S_c$ on the circumcircle of ABC, which is also the incenter of $A'B'C'$. The conclusion to part (a) follows.

Comment. The previous argument breaks down when ABC is obtuse because some of the angles cannot be added or subtracted as needed. Indeed, in this case, we can draw triangles ABC for which the statement proposed in part (a) is not true, and the incenter of $A'B'C'$ may be inside triangle ABC and not on its circumcircle.

(b) It is well known that the Miquel point for $X \in BC$, $Y \in CA$, $Z \in AB$ such that X, Y, Z are collinear, is on the circumcircle of ABC, hence M is on the circumcircle of ABC. Assume without loss of generality (since we may cyclically permute the vertices of ABC without altering the problem) that M is on the arc AB that does not contain C. Since M is on the circumcircle of AYZ, the perpendicular bisector of AM passes through O, O_a. At the same time, denoting by T the point diametrally opposite S in the circumcircle of ABC, we have, if for example $\angle SAM$ is acute and $\angle SBM$ is obtuse,

$$\begin{aligned}\angle OSM &= 90° - \angle STM = 90° - \angle SAM = 90° - \angle O_aAM = \frac{1}{2}\angle AO_aM \\ &= 180° - \angle MO_aO,\end{aligned}$$

and so M, O, O_a, S are concyclic; moreover,

$$\begin{aligned}\angle OSM &= 90^\circ - \angle STM = \angle SBM - 90^\circ = 90^\circ - \angle O_bBM = \frac{1}{2}\angle BO_bM \\ &= \angle MO_bO,\end{aligned}$$

so M, O, O_b, S are concyclic too. Similarly, since $\angle SCM$ is either acute or obtuse, we have one of these two cases, and M, O, O_c, S are also concyclic. Hence the circumcircles of SOO_a, SOO_b, SOO_c all meet again at M, which is a point on the circumcircle of ABC. The conclusion to part (b) follows.

■

Problem 46. (IMO Shortlist 2011) Let $ABCDEF$ be a convex hexagon, all of whose sides are tangent to a circle ω with center O. Suppose that the circumcircle of triangle ACE is concentric with ω. Let J be the foot of the perpendicular from B to CD. Suppose that the perpendicular from B to DF intersects the line EO at a point K. Let L be the foot of the perpendicular from K to DE. Prove that $DJ = DL$.

Solution 1. Let us denote the points of tangency of AB, BC, CD, DE, EF, and FA to ω by R, S, T, U, V, and W, respectively. As in the previous solution, we mention that

$$AR = AW = CS = CT = EU = EV.$$

The reflection in the line BO maps R to S, therefore A to C and thus also W to T. Hence, both lines RS and WT are perpendicular to OB, and are therefore parallel. On the other hand, the lines UV and WT are not parallel, since otherwise the hexagon $ABCDEF$ would be symmetric with respect to the line BO and the lines defining the point K would coincide, which contradicts the conditions of the problem. Therefore we can consider the intersection point Z of UV and WT.

Next, recall the well-known fact that points D, F, Z are collinear. Actually, D is the pole of the line UT, F is the pole of VW, and $Z = TW \cap UV$; so all these points belong to the polar line of $TU \cap VW$.

Now, we put O at the origin, and identify each point (say X) with the vector $\overrightarrow{OX}$. So, from now on all the products of points refer to the scalar products of the corresponding vectors.

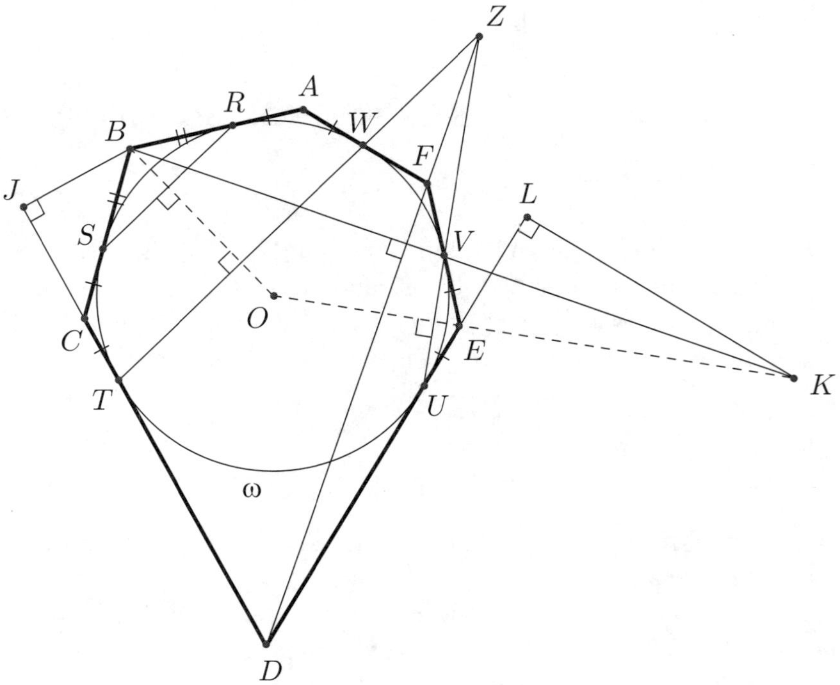

Since $OK \perp UZ$ and $OB \perp TZ$, we have $K \cdot (Z - U) = 0 = B \cdot (Z - T)$. Next, the condition $BK \perp DZ$ can be written as $K \cdot (D - Z) = B \cdot (D - Z)$. Adding these two equalities we get

$$K \cdot (D - U) = B \cdot (D - T).$$

By symmetry, we have $D \cdot (D - U) = D \cdot (D - T)$. Subtracting this from the previous equation, we obtain $(K - D) \cdot (D - U) = (B - D) \cdot (D - T)$ and rewrite it in vector form as

$$\overrightarrow{DK} \cdot \overrightarrow{UD} = \overrightarrow{DB} \cdot \overrightarrow{TD}.$$

Finally, projecting the vectors $\overrightarrow{DK}$ and $\overrightarrow{DB}$ onto the lines UD and TD respectively, we can rewrite this equality in terms of segment lengths as $DL \cdot UD = DJ \cdot TD$, thus $DL = DJ$, which completes the proof.

Solution 2. Since ω and the circumcircle of triangle ACE are concentric, the tangents from A, C, and E to ω have equal lengths; that means that $AB = BC$, $CD = DE$, and $EF = FA$. Moreover, we have $\angle BCD = \angle DEF = \angle FAB$.

Consider the rotation around point D mapping C to E; let B' and L' be the images of the points B and J, respectively, under this rotation. Then one has $DJ = DL'$ and $B'L' \perp DE$: moreover, the triangles $B'ED$ and BCD

are congruent. Since $\angle DEO < 90^\circ$, the lines EO and $B'L'$ intersect at some point K'. We intend to prove that $K'B \perp DF$; this would imply that $K = K'$, therefore $L = L'$, which proves the problem's statement.

Analogously, consider the rotation around F mapping A to E; let B'' be the image of B under this rotation. Then, the triangles FAB and FEB'' are congruent. We have $EB'' = AB = BC = EB'$ and $\angle FEB'' = \angle FAB = \angle BCD = \angle DEB'$, so the points B' and B'' are symmetrical with respect to the angle bisector EO of $\angle DEF$. So, from $K'B' \perp DE$ we get $K'B'' \perp EF$. From these two relations, we use Theorem 3 to get

$$K'D^2 - K'E^2 = B'D^2 - B'E^2 \text{ and } K'E^2 - K'F^2 = B''E^2 - B''F^2.$$

Adding these equalities and taking into account that $B'E = B''E$ we obtain

$$K'D^2 - K'F^2 = B'D^2 - B''F^2 = BD^2 - BF^2,$$

which means exactly that $K'B \perp DF$. This completes the proof.

■

Comment. There are several variations of the latter solution. For instance, let us consider the intersection point M of the lines BJ and OC. Define the point K' as the reflection of M in the line DO. Then one has

$$DK'^2 - DB^2 = DM^2 - DB^2 = CM^2 - CB^2.$$

Next, consider the rotation around O which maps CM to EK'. Let P be the image of B under this rotation; so P lies on ED. Then, $EF \perp K'P$, so

$$CM^2 - CB^2 = EK'^2 - EP^2 = FK'^2 - FP^2 = FK'^2 - FB^2,$$

since triangles FEP and FAB are congruent.

Problem 47. (Nairi Sedrakyan, International Zhautykov Olympiad 2010) A non-isosceles acute triangle ABC is given. Let O, I, H be the circumcenter, the incenter, and the orthocenter of the triangle ABC, respectively. Prove that $\angle OIH > 135^\circ$.

Solution. It is (relatively) well known that

$$OI^2 = R(R-2r);\ OH^2 = 9R^2-(a^2+b^2+c^2);\ IH^2 = 4R^2-2r^2-\frac{a^2+b^2+c^2}{2},$$

where, as usual, R, r denote the circumradius and inradius of ABC with side lengths a, b, c. We refer to [51] for a complete treatment of these type of formulas for the distances between important triangle centers. Here, we shall just note that first two relations may be found through computing the powers of I and H with respect to the circumcircle, the second one requiring a little more algebra using the Law of Cosines and the expressions for the area of a triangle. The third one may be deduced from the first two, by applying the formula for the (square of the) median in triangle OIH, since the nine-point center N is the midpoint of OH and $IN = R - \frac{r}{2}$, because the incircle and nine-point circle are tangent at the Feuerbach point (recall Feuerbach's theorem from Problem 20). Note that

$$OH^2 - IH^2 - OI^2 = IH^2 + 2r(R - 2r) > 0$$

because $R > 2r$ and $IH^2 > 0$, approaching equality when ABC approaches an equilateral triangle, or OIH will be obtuse at I.

We may also write after some algebra

$$a^2 + b^2 + c^2 = 8R^2 + 8R^2 \cos A \cos B \cos C,$$

or since ABC is acute, $a^2 + b^2 + c^2 = 8R^2 + 4\delta$ for some $\delta > 0$. Therefore, $OH^2 = R^2 - 4\delta$ And $IH^2 = 2r^2 - 2\delta$, or

$$\frac{OH^2 - IH^2 - OI^2}{2OI \cdot IH} = \frac{Rr - r^2 - \delta}{\sqrt{2}\sqrt{r^2 - \delta}\sqrt{R(R - 2r)}}.$$

Assume that this quantity is smaller than or equal to $\frac{1}{\sqrt{2}}$. Then,

$$\delta^2 + (R^2 + 2r^2)\delta + r^4 < 0,$$

which is clearly impossible. Hence $\cos \angle OIH < -\frac{1}{\sqrt{2}} = \cos 135°$, and $\angle OIH > 135°$, as claimed.

■

Problem 48. (Cyril Parry, American Mathematical Monthly) Let ABC be a triangle with circumcenter O and orthocenter H. Parallel lines α, β, γ are drawn through the vertices A, B, C, respectively. Let α', β', γ' be the reflections of α, β, γ in the sides BC, CA, AB, respectively. Prove that these reflections are concurrent if and only if α, β, γ are parallel to the Euler line OH of triangle ABC.

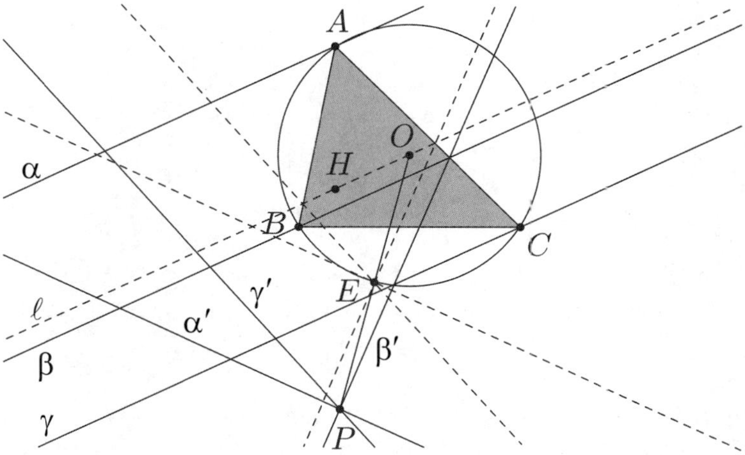

Solution. We begin with a lemma that is extremely useful when dealing with reflections of lines with respect to the sides of a certain triangle.

Collings' Theorem. Let ABC be a triangle with orthocenter H and let ℓ be a line passing through H. Prove that the reflections of ℓ across the sides of ABC are concurrent on the circumcircle of ABC.

Proof. Without loss of generality, assume that $\angle A < 90^\circ$. Let the line ℓ intersect BC at X, CA at Y, AB at Z so that Y, Z lie inside the sides, whereas X is on the extension. Let ℓ_A, ℓ_B, ℓ_C be the reflections of ℓ across BC, CA, AB, respectively. Notice that is enough to show that that the lines ℓ_B and ℓ_C meet on the circumcircle of ABC (since then we could do the same thing for the other pairs).

To see this, let P be the intersection of ℓ_B and ℓ_C. Furthermore, let H_b and H_c be the reflections of H across CA and AB, respectively. By Epsilon 18, these points lie on the circumcircle of ABC. Moreover, these points need to lie on the reflections ℓ_B and ℓ_C of ℓ (since H lies on ℓ. Now, let us prove that BXH_bH_c is cyclic. On one hand, we have that angle

$$\begin{aligned}
\angle H_bPH_c &= 180^\circ - \angle ZYP - \angle YZP \\
&= 180^\circ - (180^\circ - \angle ZYH_b) - (180^\circ - \angle YZH_c) \\
&= 180^\circ - (180^\circ - 2\angle ZYA) - (180^\circ - 2\angle YZA) \\
&= 2(\angle ZYA + \angle YZA) - 180^\circ \\
&= 2(180^\circ - \angle A) - 180^\circ \\
&= 180^\circ - 2\angle A.
\end{aligned}$$

On the other hand, note that

$$\begin{aligned}\angle H_cBH_b &= \angle H_cBA + \angle ABH_b \\ &= \angle H_cCA + \angle ABH_b \\ &= (90^\circ - \angle A) + (90^\circ - \angle A) \\ &= 180^\circ - 2\angle A.\end{aligned}$$

Hence, we conclude that $\angle H_bPH_c = \angle H_cBH_b$, and so quadrilateral BPH_bH_c is cyclic, i.e., P lies on the circumcircle of ABC. This completes the proof.

□

Returning to our problem, let $A_1B_1C_1$ be the image of ABC under the homothety $\mathcal{H}(O,2)$ with center O and (positive) ratio 2. The orthocenter H_1 of $A_1B_1C_1$ is the reflection of O in H, and is on the Euler line of triangle ABC.

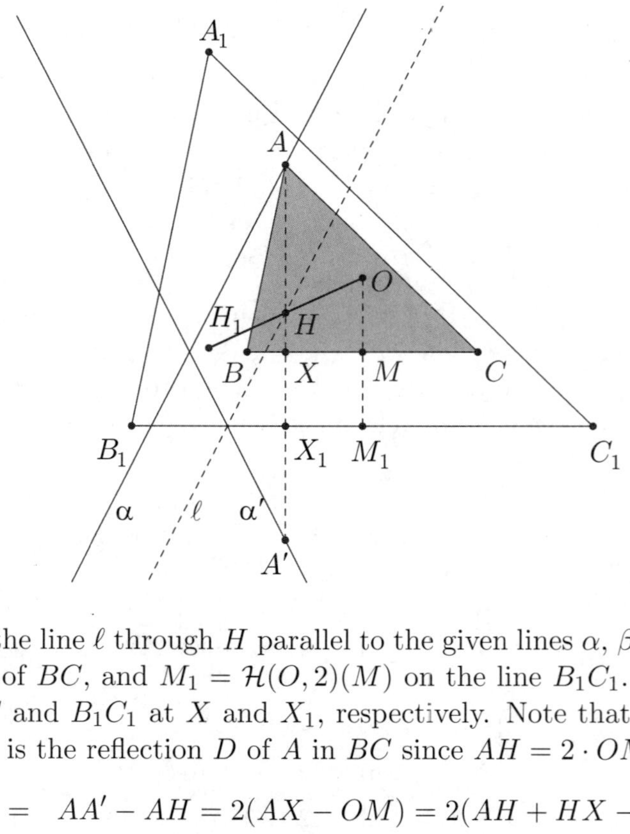

Consider the line ℓ through H parallel to the given lines α, β, γ. Let M be the midpoint of BC, and $M_1 = \mathcal{H}(O,2)(M)$ on the line B_1C_1. The line AH intersects BC and B_1C_1 at X and X_1, respectively. Note that the reflection of H in B_1C_1 is the reflection D of A in BC since $AH = 2\cdot OM$ and

$$\begin{aligned}HA' &= AA' - AH = 2(AX - OM) = 2(AH + HX - OM) \\ &= 2(HX + OM) = 2(HX + XX_1) = 2HX_1.\end{aligned}$$

Therefore, α' coincides with the reflection of ℓ in the side B_1C_1. Similarly, β' and γ' coincide with the reflections of ℓ in C_1A_1 and A_1B_1. By Colling's theorem, the lines α', β', γ' are concurrent if and only if ℓ passes through the orthocenter H_1. Since H also lies on ℓ, this is the case when ℓ is the Euler line of triangle ABC, which is also the Euler line of triangle $A_1B_1C_1$. This completes the proof.

■

Problem 49. (Luis Gonzalez and Cosmin Pohoata, Mathematical Reflections) Let ABC be a triangle with circumcircle Γ and nine-point circle γ. Let X be a point on γ, and let Y, Z be on Γ so that the midpoints of segments XY and XZ are on γ.

(a) Prove that the midpoint of YZ is on γ.

(b) Find the locus of the symmedian point of triangle XYZ, as X moves along Γ.

Solution. (a) We begin with an important property of the nine-point center of a triangle.

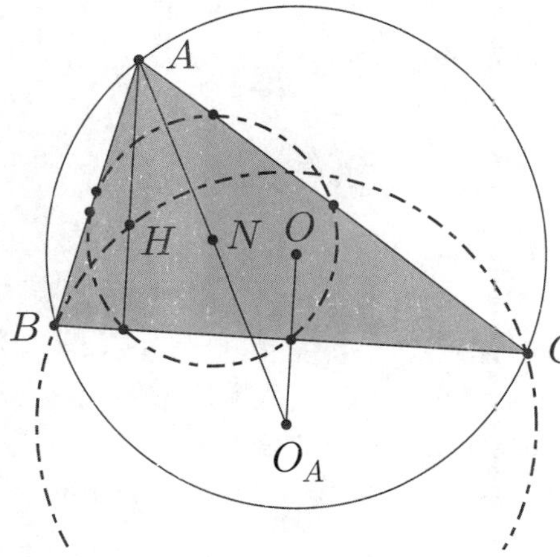

Lemma. Let ABC be a triangle with circumcenter O and nine-point center N. If O' is the reflection of O across BC, the points A, N, O' are collinear, and moreover $NA = NO'$.

Proof. The idea is to note that the reflection O' of O across BC is the circumcenter of triangle BHC, where H is the orthocenter of triangle ABC. Indeed, the circumcircles of triangles BHC and ABC are congruent (this happens, for example, because the circumcircle of BHC passes through the reflection A' of A across BC); but $O'B = O'C = OB = OC = R$, where R is the circumradius of ABC (and thus of BHC). Hence, O' is the circumcenter of BHC. Now, the collinearity of A, N, O' follows immediately, since the nine-point circle of

ABC is nothing but the homothetic image of the circumcircle of BHC under $\mathcal{H}(A, \frac{1}{2})$. Thus the Lemma is proven.

□

Now, returning to the problem, just note that the way to justify the existence of Y, Z on Γ so that the midpoints of XY and XZ are on γ is as follows: construct the homothetic image of γ under $\mathcal{H}(X, 2)$, call it γ', and then take the intersection points Y, Z of γ' and Γ. By construction, the center O' of γ' and the circumcenter O of ABC are symmetric with respect to YZ. Thus, by the above Lemma, the nine-point center N' of XYZ coincides with the midpoint of XO'. But N is the midpoint of XO'; so $N = N'$, and we are done. Triangles ABC and XYZ have the same nine-point center and share the same circumcircle. This means that they have the same nine-point circle, and so the midpoint of YZ lies on γ. This settles our problem.

(b) By part (a), all triangles XYZ have the same circumcenter O and the same nine-point center N. Since N is the midpoint of OH, where H is the orthocenter, then all triangles XYZ also have the same orthocenter H. Denote by x, y, z the respective lengths of YZ, ZX, XY. It is relatively well known (or easily found by computing the power of the orthocenter with respect to the circumcircle) that

$$OH = \sqrt{9R^2 - (x^2 + y^2 + z^2)}$$

$$= \frac{\sqrt{x^6 + y^6 + z^6 - x^4y^2 - x^4z^2 - y^4z^2 - y^4x^2 - z^4x^2 - z^4y^2 + 3x^2y^2z^2}}{4S},$$

where S is the area of XYZ. Relations expressing KO, KH as a function of the side lengths and area of XYZ are also relatively well known, and can be found for example under the entry for the symmedian point in http://mathworld.wolfram.com. Using these relations and after some algebra, we can find

$$KO^2 = R^2 - \frac{48R^2[ABC]^2}{(x^2 + y^2 + z^2)^2},$$

$$KH^2 = \frac{8(3R^2 - OH^2)S^2}{(x^2 + y^2 + z^2)^2} - \frac{R^2 - OH^2}{2}.$$

Consider now a cartesian coordinate system (α, β) such that $O \equiv (0, 0)$, $H \equiv (OH, 0)$. It follows that K satisfies

$$\alpha^2 + \beta^2 = KO^2, \qquad (\alpha - OH)^2 + \beta^2 = KH^2,$$

yielding

$$\alpha = \frac{OH^2 + KO^2 - KH^2}{2OH}, \qquad \beta^2 = KH^2 - (\alpha - OH)^2.$$

Substitution of the previous expressions for KO, KH produces

$$\alpha = \frac{3R^2 + OH^2}{4OH} - \frac{4S^2}{(9R^2 - OH^2)OH}.$$

Therefore

$$\left(\alpha - \frac{6R^2 \cdot OH}{9R^2 - OH^2}\right)^2 + \beta^2 = \frac{\left(27R^4 - 18R^2OH^2 - OH^4 - 16S^2\right)^2}{16(9R^2 - OH^2)^2OH^2}$$

$$+KH^2 - \frac{\left(27R^4 - 30R^2OH^2 + 3OH^4 - 16S^2\right)^2}{16(9R^2 - OH^2)^2OH^2}$$

$$= \frac{\left(27R^4 - 24R^2OH^2 + OH^4\right)\left(3R^2 - OH^2\right)}{2(9R^2 - OH^2)^2} - \frac{R^2 - OH^2}{2} = \left(\frac{2R \cdot OH^2}{9R^2 - OH^2}\right)^2,$$

which implies that the locus of the symmedian point is a circle, with center on ray OH, at a distance $\frac{6R^2 \cdot OH}{9R^2 - OH^2}$ from O, and with radius $\frac{2R \cdot OH^2}{9R^2 - OH^2}$. This completes the proof.

■

Comment. Part (a) also appeared in a Romanian IMO team selection test in 2013. It is also reminiscent of the celebrated

Poncelet's Porism. Suppose $\omega(I, r)$ is a circle lying inside circle $\Gamma(O, R)$ such that $R^2 - OI^2 = 2Rr$. Furthermore, let A be a point on Γ, and let B, C be the points on Γ such that lines AB and AC are tangent to ω. Then, BC is also tangent to ω.

This has a very easy proof which a lot of textbooks omit for some reason. We shall include it here for the reader's convenience.

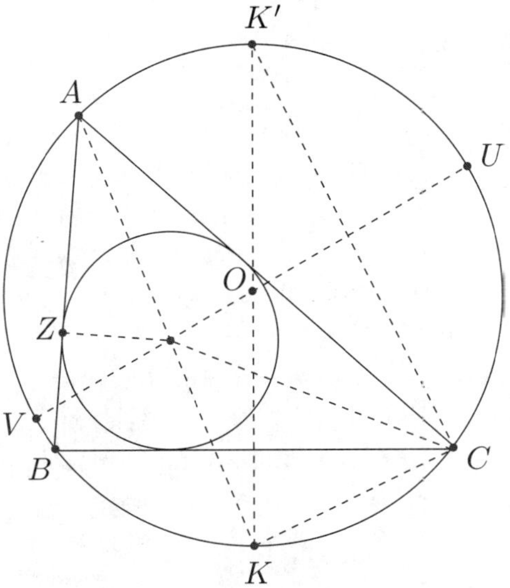

Proof. Let K be midpoint of the small arc BC, K' its antipode with respect to Γ, Z the tangency point of AB with ω, and U, V the intersections of OI with Γ, as in the diagram above. Triangles AIZ and CKK' are clearly similar; thus

$$\frac{IZ}{KK'} = \frac{AI}{CK}, \quad \text{i.e.,} \quad AI \cdot CK = 2Rr.$$

But $R^2 - OI^2$ is the power of I with respect to Γ, hence

$$AI \cdot CK = 2Rr = R^2 - OI^2 = AI \cdot IK.$$

Therefore, $CK = IK$, and so I is the incenter of triangle ABC, by the well-known property that the midpoint of the arc BC, not containing A, is the circumcenter of triangle BIC. This proves Poncelet's porism.

As an exercise, try to prove part (a) of the initial problem by reducing it to Poncelet's theorem.

Problem 50. (Cosmin Pohoata, IMO 2012 Shortlist) Let ABC be a triangle with circumcircle ω and ℓ a line not intersecting ω. Denote by P the foot of the perpendicular from the center of ω to ℓ. The side-lines BC, CA, AB intersect ℓ at the points X, Y, Z different from P. Prove the circumcircles of triangles AXP, BYP and CZP have a common point different from P or are mutually tangent at P.

Solution. Let ω_A, ω_B, ω_C and ω be the circumcircles of triangles AXP, BYP, CZP and ABC respectively. The strategy of the proof is to construct a point Q with the same power with respect to the four circles. Then each of P and Q has the same power with respect to ω_A, ω_B, ω_C and hence the three circles are coaxial. In other words, they have another common point P' or the three of them are tangent at P.

We first give a description of the point Q. Let $A' \neq A$ be the second intersection of ω and ω_A; define B' and C' analogously. We claim that AA', BB' and CC' have a common point. Once this claim is established, the point just constructed will be on the radical axes of the three pairs of circles $\{\omega, \omega_A\}$, $\{\omega, \omega_B\}$, $\{\omega, \omega_C\}$. Hence it will have the same power with respect to $\omega, \omega_A, \omega_B, \omega_C$.

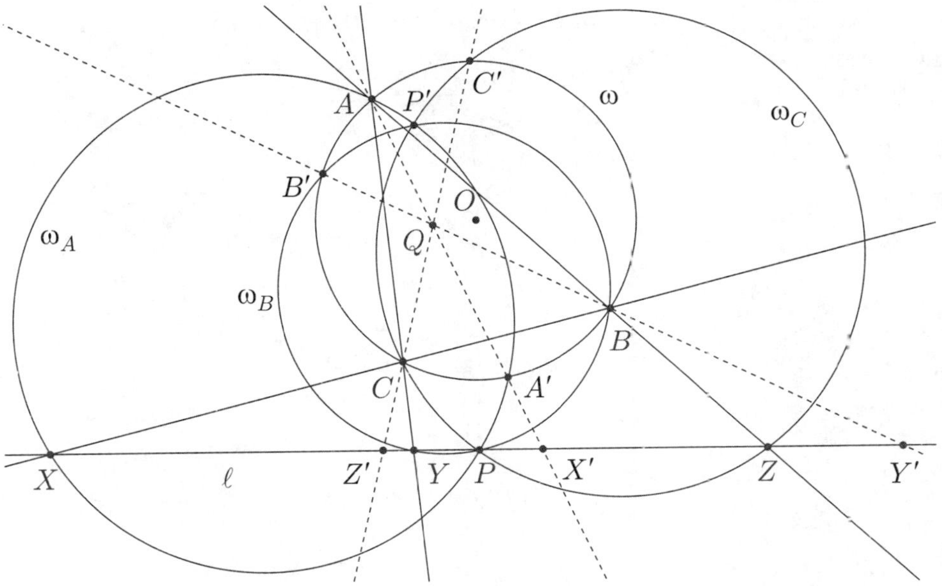

We proceed to prove that AA', BB' and CC' intersect at one point. Let R be the circumradius of ABC. Define the points X', Y', Z' as the intersections of AA', BB', CC' with ℓ. Observe that X', Y', Z' do exist. If AA' is parallel to ℓ then ω_A is tangent to ℓ; hence $X = P$, which is a contradiction. Similarly, BB' and CC' are not parallel to ℓ.

From the powers of the point X' with respect to the circles ω_A and ω we get

$$X'P \cdot (X'P + PX) = X'P \cdot X'X = X'A' \cdot X'A = X'O^2 - R^2,$$

hence

$$X'P \cdot PX = X'O^2 - R^2 - X'P^2 = OP^2 - R^2.$$

Analogous relations hold for the points Y' and Z', so we obtain

$$X'P \cdot PX = Y'P \cdot PY = Z'P \cdot PZ = OP^2 - R^2 = k^2.$$

Call this chain of equalities (1) for reference purposes. In these computations all segments are regarded as directed segments. We keep the same convention for the sequel.

We prove that the lines AA', BB', CC' intersect at one point by Ceva's theorem. To avoid distracting remarks we interpret everything projectively, i.e., whenever two lines are parallel they meet at a point on the line at infinity.

Let U, V, W be the intersections of AA', BB', CC' with BC, CA, AB, respectively. The idea is that although it is difficult to calculate the ratio $\frac{BU}{CU}$, it is easier to deal with the cross-ratio $\frac{BU}{CU} / \frac{BX}{CX}$ because we can send it to the line ℓ. With this in mind we apply Menelaus' theorem to the triangle ABC and obtain

$$\frac{BX}{CX} \cdot \frac{CY}{AY} \cdot \frac{AZ}{BZ} = 1.$$

Hence Ceva's ratio can be expressed as

$$\frac{BU}{CU} \cdot \frac{CV}{AV} \cdot \frac{AW}{BW} = \frac{BU}{CU} : \frac{BX}{CX} \cdot \frac{CV}{AV} : \frac{CY}{AY} \cdot \frac{AW}{BW} : \frac{AZ}{BZ}.$$

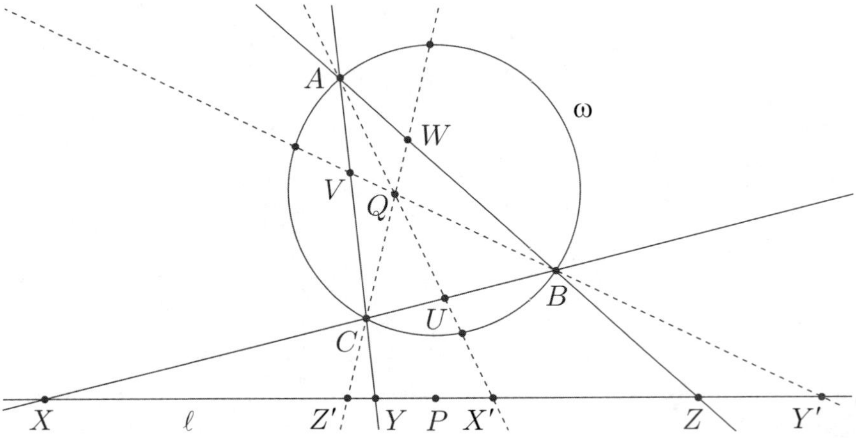

Project the line BC to ℓ from A. The cross-ratio between BC and UX equals the cross-ratio between ZY and $X'X$. Repeating the same argument with the lines CA and AB gives

$$\frac{BU}{CU} \cdot \frac{CV}{AV} \cdot \frac{AW}{BW} = \frac{ZX'}{YX'} : \frac{ZX}{YX} \cdot \frac{XY'}{ZY'} : \frac{XY}{ZY} \cdot \frac{YZ'}{XZ'} : \frac{YZ}{XZ}$$

and hence

$$\frac{BU}{CU} \cdot \frac{CV}{AV} \cdot \frac{AW}{BW} = (-1) \cdot \frac{ZX'}{YX'} \cdot \frac{XY'}{ZY'} \cdot \frac{YZ'}{XZ'}.$$

The equations (1) reduce the problem to a straightforward computation on the line ℓ. For instance, the transformation $t \to -k^2/t$ preserves cross-ratio and interchanges the points X, Y, Z with the points X', Y', Z'. Then

$$\frac{BU}{CU} \cdot \frac{CV}{AV} \cdot \frac{AW}{BW} = (-1) \cdot \frac{ZX'}{YX'} : \frac{ZZ'}{YZ'} \cdot \frac{XY'}{ZY'} : \frac{ZX'}{ZZ'} = -1.$$

We proved that Ceva's ratio equals -1, so AA', BB', CC' intersect at one point Q.

Comment 1. There is a nice projective argument to prove that AX', BY', CZ' intersect at one point. Suppose that ℓ and ω intersect at a pair of complex conjugate points D and E. Consider a projective transformation that takes D

and E to $[i,1,0]$ and $[-i,1,0]$. Then ℓ is the line at infinity, and ω is a conic through the special points $[i,1,0]$ and $[-i,1,0]$, hence it is a circle. So one can assume that AX, BY, CZ are parallel to BC, CA, AB. The involution on ℓ taking X, Y, Z to X', Y', Z' and leaving D, E fixed is the involution changing each direction to its perpendicular one. Hence AX, BY, CZ are also perpendicular to AX', BY', CZ'.

It follows from the above that AX', BY', CZ' intersect at the orthocenter of triangle ABC.

Comment 2. The restriction that the line ℓ does not intersect the circumcircle ω is unnecessary. The proof above works in general. In case ℓ intersects ω at D and E, point P is the midpoint of DE and some equations can be interpreted differently. For instance

$$X'P \cdot X'X = X'A' \cdot X'A = X'D \cdot X'E$$

and hence the pairs $X'X$ and DE are harmonic conjugates. This means that X', Y', Z' are the harmonic conjugates of X, Y, Z with respect to the segment DE.

Solution 2. First we prove that there is an inversion in space that takes ℓ and ω to parallel circles on a sphere. Let QR be the diameter of ω whose extension beyond Q passes through P. Let Π be the plane containing these objects. In space, choose a point O such that the line QO is perpendicular to Π and $\angle POR = 90°$, and apply an inversion with center O and arbitrary power. For any object $\mathcal{T}$ denote by $\mathcal{T}'$ the image of $\mathcal{T}$ under this inversion.

The inversion takes the plane Π to a sphere Π'. The lines in Π are taken to circles through O, and the circles in Π are also taken to circles on Π'.

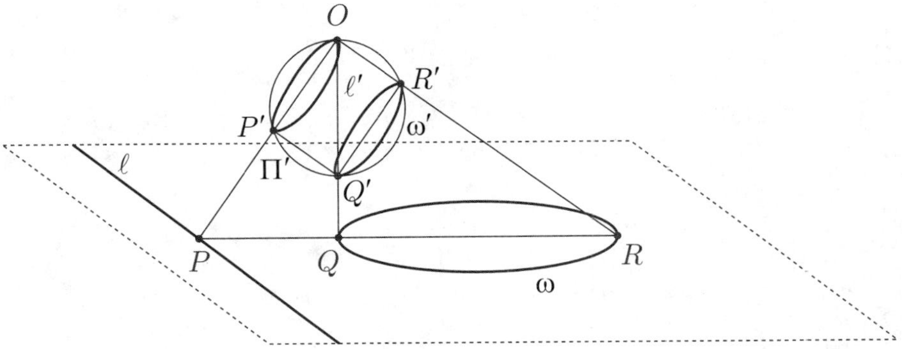

Since the line ℓ and the circles ω are perpendicular to the plane OPQ, the circles ℓ' and ω' also are perpendicular to this plane. Hence, the planes of the circles ℓ' and ω' are parallel.

Now consider the circles $A'X'P'$, $B'Y'P'$ and $C'Z'P'$. We want to prove that either they have a common point (on Π'), different from P', or they are tangent to each other. The point X' is the second intersection of the circles $B'C'O$ and ℓ', other than O. Hence, the lines OX' and $B'C'$ are coplanar. Moreover, they lie in the parallel planes of ℓ' and ω'. Therefore, OX' and $B'C'$ are parallel. Analogously, OY' and OZ' are parallel to $A'C'$ and $A'B'$.

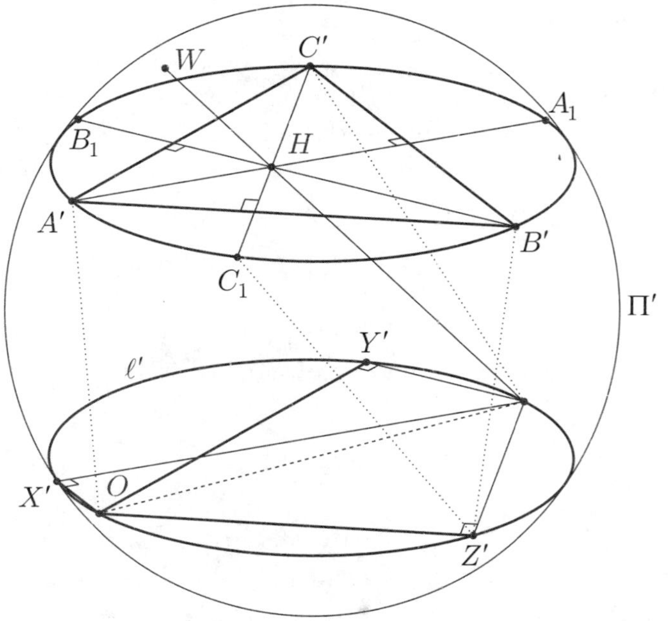

Let A_1 be the second intersection of the circles $A'X'P'$ and ω' other than A'. The segments $A'A_1$ and $P'X'$ are coplanar, and therefore parallel. Now we know that $B'C'$ and $A'A_1$ are parallel to OX' and $X'P'$ respectively, but these two segments are perpendicular because OP' is a diameter of ℓ'. We found that $A'A_1$ and $B'C'$ are perpendicular, hence $A'A_1$ is the altitude in the triangle $A'B'C'$, starting from A. Analogously, let B_1 and C_1 be the second intersections of ω' with the circles $B'P'Y'$ and $C'P'Z'$, other than B' and C' respectively. Then $B'B_1$ and $C'C_1$ are the two other altitudes in the triangle $A'B'C'$.

Let H be the orthocenter of triangle $A'B'C'$ and let W be the second intersection of the line $P'H$ with the sphere Π'. other than P'. The point W lies on the sphere Π', in the plane of the circle $A'P'X'$, so W lies on the circle $A'P'X'$. Similarly, W lies on the circles $B'P'Y'$ and $C'P'Z'$ as well; indeed W is the second common point of the three circles. If the line $P'H$ is tangent to the sphere then W coincides with P', and $P'H$ is the common tangent of the three circles.

■

Problem 51. (Cosmin Pohoata, College Mathematics Journal) Let ABC be an acute-angled triangle and let H be its orthocenter. Let t_a, t_b, t_c be the inradii of triangles HBC, HCA, and HAB, respectively. Prove that

$$t_a + t_b + t_c \leq (6\sqrt{3} - 9)r,$$

where r is the inradius of triangle ABC.

Solution. At the heart of our proof stand two very different results that come together to create magic. The first one is the well-known

Japanese Theorem. Let $ABCD$ be a cyclic quadrilateral and let r_a, r_b, r_c, r_d be the inradii of triangles BCD, CDA, DAB, ABC, respectively. Then,

$$r_a + r_c = r_b + r_d.$$

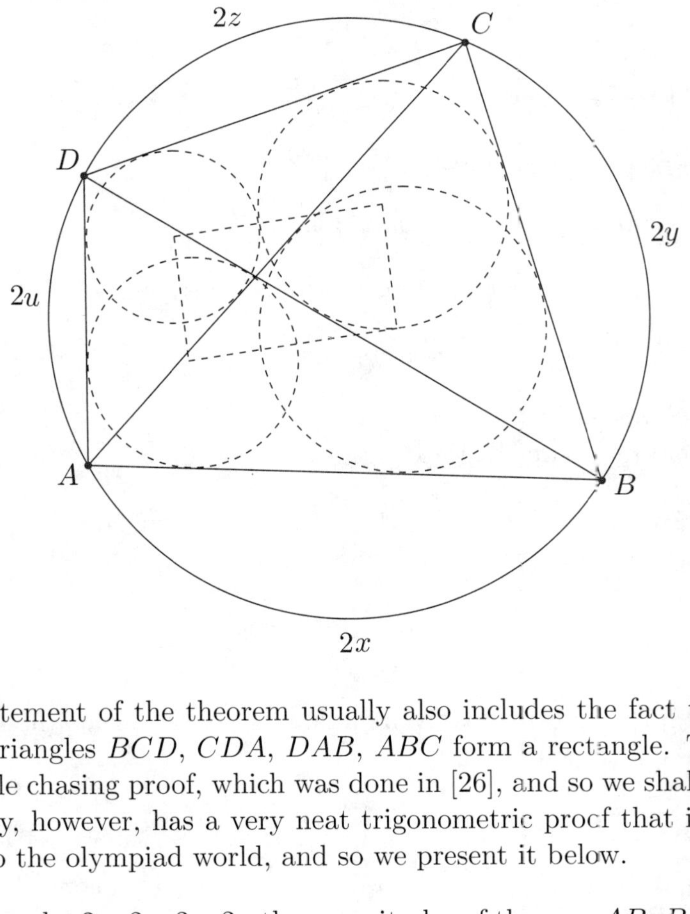

The statement of the theorem usually also includes the fact that the incenters of triangles BCD, CDA, DAB, ABC form a rectangle. This admits a short angle chasing proof, which was done in [26], and so we shall omit here. The identity, however, has a very neat trigonometric proof that is seemingly unknown to the olympiad world, and so we present it below.

Proof. Denote by $2x$, $2y$, $2z$, $2u$ the magnitudes of the arcs AB, BD, CD, DA of the circumcircle of quadrilateral $ABCD$. We thus have that $x + y + z + u = 180°$. Furthermore, let R be the circumradius of our cyclic quadrilateral.

Since

$$r_a = 4R \sin\frac{x}{2}\sin\frac{u}{2}\sin\frac{y+z}{2}, \qquad r_c = 4R \sin\frac{y}{2}\sin\frac{z}{2}\sin\frac{x+u}{2},$$

we get

$$\begin{aligned} r_a + r_c &= 2R\left(\cos\frac{x-u}{2} - \cos\frac{x+u}{2}\right)\sin\frac{y+z}{2} \\ &\quad +2R\left(\cos\frac{y-z}{2} - \cos\frac{y+z}{2}\right)\sin\frac{x+u}{2}. \end{aligned}$$

Now because $(y+z)/2 = 90^\circ - (x+u)/2$, this means that

$$\sin\frac{y+z}{2} = \cos\frac{x+u}{2} \quad \text{and} \quad \sin\frac{x+u}{2} = \cos\frac{y+z}{2},$$

from which we deduce that

$$\begin{aligned} r_a + r_c &= 2R\left(\cos\frac{x-u}{2}\cos\frac{x+u}{2} - \cos^2\frac{x+u}{2} + \cos\frac{y-z}{2}\cos\frac{y+z}{2} - \cos^2\frac{y+z}{2}\right) \\ &= R\left(\cos x + \cos y + \cos z + \cos u - 2\right). \end{aligned}$$

Viewing the symmetry in x, y, z, u, it follows that

$$r_a + r_c = r_b + r_d = R\left(\cos x + \cos y + \cos z + \cos u - 2\right).$$

This proves the claim.

□

The second result we shall use is the very powerful

Blundon's Inequality. Let ABC be a scalene triangle and let s, r, R be its semiperimeter, inradius, and its circumradius, respectively. Then,

$$s \le 2R + (3\sqrt{3} - 4)r.$$

Proof. First, note that any positive real numbers x, y, z such that $x+y+z = xyz$ satisfy the inequality

$$(x-1)(y-1)(z-1) \le 6\sqrt{3} - 10.$$

Indeed, since the numbers are positive, from the given condition it follows immediately that $x < xyz$, i.e., $yz > 1$, and similarly $xy > 1$ and $xz > 1$, so it is impossible fo more than one of the numbers to be less than or equal to 1. Because if a number is less than or equal to 1 and two are greater than

1 the inequality is obviously true (the product from the left-hand side being nonpositive), we still have to consider the case when $x, y, z > 1$. Then the numbers $u = x - 1, v = y - 1$ and $w = z - 1$ are positive and, replacing $x = u + 1, y = v + 1, w = z + 1$ in the condition from the hypothesis, one gets

$$uvw + uv + uw + vw = 2.$$

By the AM-GM inequality,

$$uvw + 3\sqrt[3]{u^2v^2w^2} \leq uvw + uv + uw + vw = 2,$$

and hence for $t = \sqrt[3]{uvw}$ we have $t^3 + 3t^2 - 2 \leq 0$, or equivalently

$$(t + 1)(t + 1 + \sqrt{3})(t + 1 - \sqrt{3}) \leq 0.$$

We conclude that $t \leq \sqrt{3} - 1$, which gives

$$(x - 1)(y - 1)(z - 1) \leq 6\sqrt{3} - 10.$$

The equality occurs when $x = y = z = \sqrt{3}$.

Returning to Blundon's inequality, we refer to the well known formulas

$$\cot\frac{A}{2} = \sqrt{\frac{s(s-a)}{(s-b)(s-c)}}, \quad \cot\frac{B}{2} = \sqrt{\frac{s(s-b)}{(s-c)(s-a)}}, \quad \cot\frac{C}{2} = \sqrt{\frac{s(s-c)}{(s-a)(s-b)}},$$

to deduce that

$$\sum_{\text{cyclic}} \cot\frac{A}{2} = \prod_{\text{cyclic}} \cot\frac{A}{2} = \frac{s}{r},$$

and

$$\sum_{\text{cyclic}} \cot\frac{A}{2}\cot\frac{B}{2} = \sum_{\text{cyclic}} \frac{s}{s-a} = \frac{4R + r}{r}.$$

In this case, by applying the result from the above paragraph for the positive numbers $x = \cot\frac{A}{2}$, $y = \cot\frac{B}{2}$ and $z = \cot\frac{C}{2}$, it follows that

$$\left(\cot\frac{A}{2} - 1\right)\left(\cot\frac{B}{2} - 1\right)\left(\cot\frac{C}{2} - 1\right) \leq 6\sqrt{3} - 10,$$

and therefore

$$2\prod_{\text{cyclic}} \cot\frac{A}{2} - \left(\sum_{\text{cyclic}} \cot\frac{A}{2}\cot\frac{B}{2}\right) \leq 6\sqrt{3} - 9.$$

This rewrites as

$$\frac{2s}{r} - \frac{4R+r}{r} \le 6\sqrt{3} - 9,$$

and thus

$$s \le 2R + (3\sqrt{3} - 4)r.$$

The equality occurs if and only if $\cot\frac{A}{2} = \cot\frac{B}{2} = \cot\frac{C}{2}$, i.e., when the triangle ABC is equilateral. This completes the proof of Blundon's Inequality.

□

We finally return to our problem. Let A' be the antipode of A with respect to triangle ABC. It is well known that A' is the reflection of the orthocenter H with respect to the midpoint of the side BC. In this case, the triangles HBC and $A'CB$ are congruent, and so their inradii are equal.

In what follows, we shall denote by r_{XYZ}, s_{XYZ} the inradius and semiperimeter of triangle XYZ, respectively. Thus, we just proved that $t_a = r_{A'BC}$. According to the Japanese theorem, applied to the cyclic quadrilateral $ABA'C$, we have that

$$r_{ABC} + r_{A'BC} = r_{ABA'} - r_{ACA'},$$

i.e.,

$$r + t_a = r_{ABA'} + r_{ACA'}.$$

Now note that triangles ABA' and ACA' are both right-angled triangles (at B and C, respectively) since AA' is a diameter of the circumcircle of ABC. Thus, we have that

$$r_{ABA'} = s_{ABA'} - AA' = \frac{AB + A'B - AA'}{2} = \frac{AB + HC - AA'}{2}.$$

Similarly,

$$r_{ACA'} = s_{ACA'} - AA' = \frac{AC + A'C - AA'}{2} = \frac{AC + HB - AA'}{2}.$$

Thus,

$$\begin{aligned} r_{ABA'} + r_{ACA'} &= \frac{AB + HC - AA'}{2} + \frac{AC + HB - AA'}{2} \\ &= \frac{AB + AC + HB + HC}{2} - 2R, \end{aligned}$$

where R is the circumradius of ABC.

Hence,

$$r + t_a = r_{ABA'} + r_{ACA'} = \frac{AB + AC + HB + HC}{2} - 2R.$$

Similarly, we deduce the analogous relations

$$r + t_b = \frac{BC + BA + HC + HA}{2} - 2R,$$

and

$$r + t_c = \frac{CA + CB + HA + HB}{2} - 2R.$$

Summing up these last three relations, we get

$$3r + t_a + t_b + t_c = 2s + HA + HB + HC - 6R.$$

But triangle ABC is acute-angled, and so

$$HA = 2R\cos A,\ HB = 2R\cos B,\ HC = 2R\cos C.$$

Therefore,

$$\begin{aligned} t_a + t_b + t_c &= 2s + 2R(\cos A + \cos B + \cos C) - 6R - 3r \\ &= 2s + 2(R + r) - 6R - 3r \\ &= 2s - 4R - r \\ &\le \left(6\sqrt{3} - 9\right) r, \end{aligned}$$

where the second equality holds because $\cos A + \cos B + \cos C = 1 + \frac{R}{r}$ and the last one holds by Blundon's inequality. This completes the proof.

■

Problem 52. (Asian Pacific Mathematical Olympiad 2010) Let ABC be an acute angled triangle satisfying the conditions $AB > BC$ and $AC > BC$. Denote by O and H the circumcenter and orthocenter, respectively, of the triangle ABC. Suppose that the circumcircle of the triangle AHC intersects the line AB at M different from A, and the circumcircle of the triangle AHB intersects the line AC at N different from A. Prove that the circumcenter of the triangle MNH lies on the line OH.

Solution. Consider the inversion Ψ with center H and power $HA \cdot HB$. By Power of Point, this sends the vertices of triangle ABC to the corresponding feet of the altitudes, say D, E, F. The sidelines BC, CA, AB of triangle ABC thus go to the circumcircles of quadrilaterals $HEAF$, $HDFB$ and $HDEC$, respectively. The circles HAB and HAC go to the lines DE and DF, and if we denote by X' the image of X under Ψ, we get that M', N' are the intersection points of the circles $(HDEC)$ and $(HDFB)$ with lines DF and DB. Hence, the statement to prove is equivalent to showing that TH is perpendicular to $M'N'$, where T is the nine-point center of ABC.

Recall that H is the incenter of the orthic triangle DEF and that T is its circumcenter. It is also easy to see that the points M' and N' are such that E lies outside the segment $M'D$ and F lies outside the segment $N'D$, and furthermore so that $EM' = FN' = EF$. To show that HT is orthogonal to $M'N'$ we plan to verify that $HM'^2 + TM'^2 - HN'^2 - TN'^2 = 0$. Indeed, we write

$$\begin{aligned} TM'^2 &= M'D \cdot M'E + \frac{TD^2}{4} = (M'E - DE)M'E + \frac{TD^2}{4} \\ &= M'E^2 - M'E \cdot DE + \frac{TD^2}{4} \\ &= EF^2 - DE \cdot EF + \frac{TD^2}{4}. \end{aligned}$$

Analogously, $TN'^2 = EF^2 - DF \cdot EF + \frac{TD^2}{4}$. Also, if H' and H'' are the orthogonal projections of H on DE and DF (note that $HH' = HH''$), we have that

$$\begin{aligned} HM'^2 &= HH'^2 + H'M'^2 \\ &= HH'^2 + (EM' - EH')^2 \\ &= HH'^2 + \left(EF - \frac{EF + ED - FD}{2}\right)^2. \end{aligned}$$

Similarly, we also get that

$$HN'^2 = HH''^2 + \left(EF - \frac{EF + FD - ED}{2}\right)^2;.$$

hence, a simple verification yields $HM'^2 + TM'^2 - HN'^2 - TN'^2 = 0$, as desired. ■

Problem 53. (Luis Gonzalez and Cosmin Pohoata, IMO 2014 Proposal) Let $ABCD$ be a rectangle and let ω be an arbitrary circle passing through vertices A and C. Let Γ_1, Γ_2 be the circles inside $ABCD$ so that Γ_1 is tangent to AB, BC, and Γ, and Γ_2 is tangent to CD, DA, and Γ. Prove that the sum of the radii of Γ_1 and Γ_2 is independent of the choice of ω.

Solution. A computational approach using Casey's theorem is possible. Nonetheless, we leave that to the reader and show you the way we created the problem. Let $r_1 \geq r_2$ be the inradii of $\Gamma_1(O_1)$ and $\Gamma_2(O_2)$, respectively. Furthermore, let Γ_1 touch AB, BC at P, Q and let Γ_2 touch CD, DA at R, S, respectively. Let the line BO_1 intersect PQ and RS at F and G and let DO_2 intersect RS at H, respectively. If δ then denotes the distance between the obviously parallel external angle bisectors of $\angle ABC$ and $\angle CDA$, we have that

$$\delta = BF + FG + HD = FG + \tfrac{\sqrt{2}}{2}(r_1 + r_2).$$

Thus, it suffices to prove that FG is constant, since δ is clearly constant.

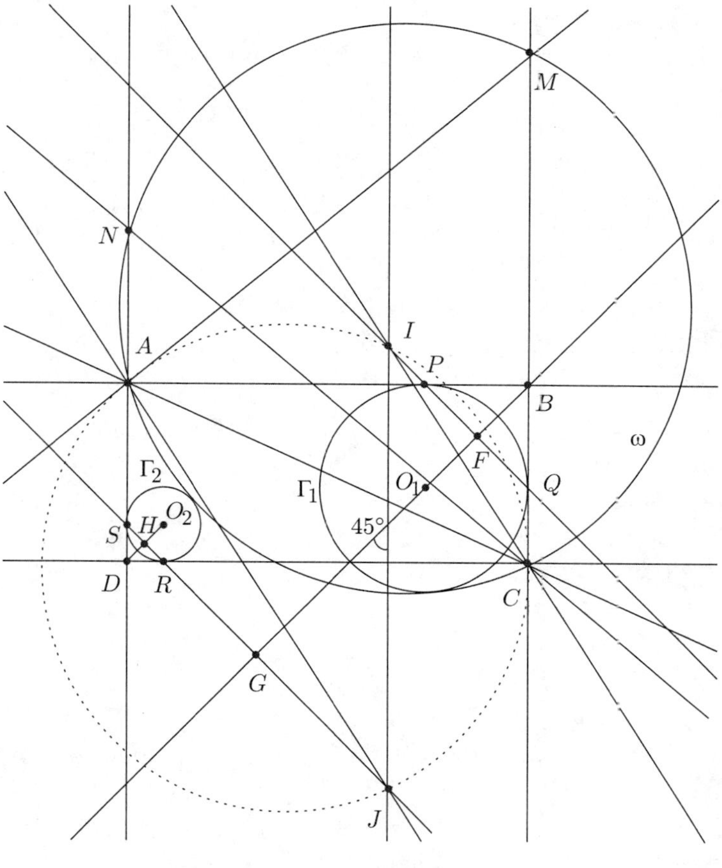

Let ω intersect the lines BC and DA again at points M and N, respectively. The circle Γ_1 thus becomes a Thebault circle for the cevian AB of triangle ACM, being internally tangent to the circumcircle ω, whereas Γ_2 is a Thebault circle for the cevian CD of triangle CAN, which has the same circumcircle ω. Sawayama's theorem, which we discussed in the proof of Problem 36, tells us now that PQ goes through the incenter I of triangle ACM and RS goes through the N-excenter J of triangle CAN. This implies that

$$\angle AIC = 90^\circ + \frac{1}{2}\angle AMC = 90^\circ + \frac{1}{2}\angle ANC = 180^\circ - \angle AJC.$$

Hence, quadrilateral $AICJ$ is cyclic - and since the bisectors CI and AJ of angles ACB and CAD are obviously parallel, $AICJ$ is an isosceles trapezoid with $IJ = AC$. Moreover,

$$\angle CIJ = \angle CAJ = \angle JAD;$$

thus $IJ \parallel AD \parallel BC$, which implies $\angle(IJ, FG) = 45°$. It follows that

$$FG = \frac{\sqrt{2}}{2} IJ = \frac{\sqrt{2}}{2} AC,$$

which is clearly a constant, as claimed. This completes the proof.

■

Problem 54. (Fedor Petrov and Cosmin Pohoata, MathLinks Contest 2008) Given is an acute-angled triangle ABC and the points A_1, B_1, C_1, which are the feet of its altitudes from A, B, C, respectively. A circle passes through A_1 and B_1 and touches the smaller arc AB of the circumcircle of ABC at point C_2. Points A_2 and B_2 are defined analogously. Prove that the lines A_1A_2, B_1B_2, C_1C_2 have a common point, which lies on the Euler line of triangle ABC.

Solution. Define the intersections $X = B_1C_1 \cap BC$, $Y = C_1A_1 \cap CA$, $Z = A_1B_1 \cap AB$, and let H be the orthocenter of triangle ABC. Since lines AA_1, BB_1, CC_1 are concurrent, the quadruple (B, A_1, C, X) is a harmonic division. In particular, this means that A_1 lies on the polar of X with respect to the circumcircle Γ of triangle ABC. We claim that this polar is in fact the line A_1A_2.

Indeed, $\angle BB_1C = \angle BC_1C = 90°$, so the quadrilateral BCB_1C_1 is cyclic. Thus, X is the radical center of the circumcircles of triangles ABC, $B_1C_1A_2$ and BCB_1, and so XA_2 is tangent to the circle Γ at A_2. This implies that A_2 lies on the polar of X with respect to Γ, and thus it follows that A_1A_2 is the polar of X. Analogously, we get that B_1B_2 is the polar of Y and C_1C_2 is the polar of Z, all with respect to Γ. For the first part of the problem, it thus suffices to show that the points X, Y and Z are collinear.

Let O and N be the circumcenters of triangles ABC and $A_1B_1C_1$, respectively. As already noted, quadrilateral BCB_1C_1 is cyclic, so $XB \cdot XC = XB_1 \cdot XC_1$, or in other words X lies on the radical axis of circles (O) and (N). Similarly, Y and Z also lie on the radical axis of circles (O) and (N). Therefore, X, Y, Z lie on a line perpendicular to ON, the Euler line of triangle ABC.

On the other hand, by pole-polar duality, the collinearity of X, Y, Z yields that A_1A_2, B_1B_2 and C_1C_2 are concurrent at the pole P of line XYZ with respect to Γ, which means that $OP \perp XYZ$. This, combined with the above, allows us to conclude that O, N and P are collinear, and so we have proved that P lies on the Euler line of ABC, as claimed.

■

Problem 55. (Lev Emelyanov and Vladimir Zajic) Let ABC be a triangle with nine-point circle ω. Prove that

(a) There exist precisely three points on the circumcircle of triangle ABC whose Simson lines with respect to the triangle ABC are tangent to ω.

(b) The triangle formed by the three points from (a) is equilateral.

Solution. (a) This problem tests the reader's familiarity with the Simson line even more than Problems 7 and 17. The key is to use Lemma from the proof of Problem 17 and the following little extra observation about the angle between the Simson lines of two points lying on the circumcircle of a triangle.

Lemma. If P and Q are two points lying on the circumcircle of ABC, then the (acute) angle between their Simson lines is equal to $\frac{1}{2}\angle POQ$.

Proof. Without loss of generality, assume that P lies on the arc BC not containing the vertex A and that Q lies on the arc CA not containing B. Let D, E be the projections of P on the sidelines BC, CA and X, Y the projections of Q on BC and CA, respectively. Note that

$$\begin{aligned}
\angle(s_P, s_Q) &= 180^\circ - \angle(s_P, AC) - \angle(AC, s_Q)\\
&= 180^\circ - \angle DEA - \angle XYC\\
&= 180^\circ - \angle DPC - \angle XQC\\
&= 180^\circ - (90^\circ - \angle PCB) - (90^\circ - \angle QCB)\\
&= \angle PCB + \angle QCB\\
&= \angle PCQ\\
&= \frac{1}{2}\angle POQ.
\end{aligned}$$

This completes the proof. □

Returning to the problem, let X be a point on the circumcircle of ABC such that s_x is tangent to the nine-point circle of ABC. Such a point exists by a standard continuity argument. Suppose that this point lies on the arc BC, not containing the vertex A. If N is the nine-point center of ABC and X' the midpoint of HX, the tangency condition rewrites as $NX' \perp s_x$ (by the Lemma of Problem 17, the midpoint X' of HX is an intersection of the Simson line of X with the nine-point circle). But N is the midpoint of HO. Therefore, $OX \perp s_x$.

Now, if another point Y lies on the same arc BC, not containing A, and satisfies $OY \perp s_y$, by the above Lemma, we get that

$$\frac{1}{2}\angle XOY = \angle(s_x, s_y) = \angle(OX, OY) = \angle XOY.$$

Hence $\angle XOY$ is the null angle, i.e., $X \equiv Y$. It follows that there are exactly three points X, Y, Z on the circumcircle (lying on the small arcs BC, CA, AB, respectively) with their Simson lines tangent to the nine-point circle. This proves (a).

(b) For the second part of the problem, we shall evaluate the angles $\angle XOY$, $\angle YOZ$ and $\angle ZOX$.

Let s'_x, s'_y be the parallels through X, Y to s_x and s_y, respectively. Denote by Q the intersection point of s'_x and s'_y. Since $OX \perp s'_x$ and $OY \perp s'_y$, the quadrilateral $OXQY$ is cyclic and so

$$\angle XOY = 180^\circ - \angle XQY = 180^\circ - \angle(s_x, s_y) = 180^\circ - \frac{1}{2}\angle XOY,$$

where the latter equality holds by the Lemma we proved. This yields $\angle XOY = 120^\circ$, and similarly, we establish that $\angle YOZ = \angle ZOX = 120^\circ$. Therefore, XYZ is equilateral.

■

Comment. By angle-chasing, one can also prove that the (acute) angle between lines YZ and BC is equal to $\frac{1}{3}|B - C|$, which is also the angle between lines $B'C'$ and BC, where $A'B'C'$ is the (equilateral) Morley triangle of ABC from Problem 27. Hence, as a matter of fact, we have that triangles XYZ is homothetic to triangle $A'B'C'$. We let the reader to fill in the details. The original problem was also given in a Romanian IMO team selection test in 2009.

Problem 56. (IMO 2011) Let ABC be an acute triangle with circumcircle ω. Let ℓ be a tangent line to ω. Let ℓ_a, ℓ_b and ℓ_c be the lines obtained by reflecting ℓ in the lines BC, CA and AB, respectively. Show that the circumcircle of the triangle determined by the lines ℓ_a, ℓ_b and ℓ_c is tangent to the circle ω.

Solution 1. Even though we generally avoid doing this throughout the book, we will use the notion of oriented angles to avoid a large case distinction. Namely, for two lines ℓ and m, we denote by $\angle(\ell, m)$ the angle by which one may rotate ℓ anticlockwise to obtain a line parallel to m. Thus, all oriented angles are considered modulo 180°.

Denote by T the point of tangency of ℓ and ω. Let $A' = \ell_b \cap \ell_c$, $B' = \ell_a \cap \ell_c$, $C' = \ell_a \cap \ell_b$. Introduce the point A'' on ω such that $TA = AA''$ ($A'' \neq T$ unless TA is a diameter). Define the points B'' and C'' in a similar way.

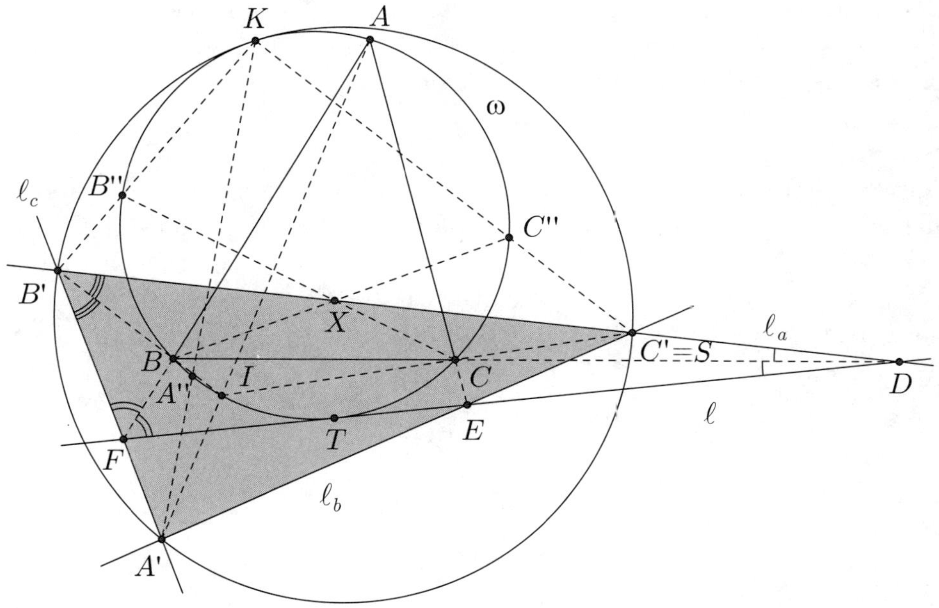

Since the points C and B are the midpoints of arcs TC'' and TB'', respectively, we have

$$\begin{aligned}\angle(\ell, B''C'') &= \angle(\ell, TC'') + \angle(TC'', B''C'')\\ &= 2\angle(\ell, TC) + 2\angle(TC'', BC'')\\ &= 2\left(\angle(\ell, TC) + \angle(TC, TB)\right)\\ &= 2\angle(\ell, BC)\\ &= \angle(\ell, \ell_a).\end{aligned}$$

It follows that ℓ_a and $B''C''$ are parallel. Similarly, $\ell_b \| A''C''$ and $\ell_c \| A''B''$. Thus, either the triangles $A'B'C'$ and $A''B''C''$ are homothetic, or they are translations of each other (i.e., their homothetic center is at infinity). Now we will prove that they are in fact homothetic, and that the center K of the homothety lies on ω. It would then follow that their circumcircles are also homothetic with respect to K and are therefore tangent at this point, as desired.

We first note that the point of intersection X of the lines $B''C$ and BC'' lies on ℓ_a. Indeed, the points X and T are actually symmetric about the line BC, since the lines CT and CB'' are symmetric about this line, as are the lines BT and BC''. Second, note that the lines BB' and CC' meet at a point that lies on the circle ω, by Problem 45, part (a). Call this point I. Furthermore, let K be the second intersection of $B'B''$ and ω. Applying Pascal's theorem to hexagon $KB''CIBC''$, we get that the points $B' = KB'' \cap IB$ and $X = B''C \cap BC''$ are collinear with the intersection point S of CI and $C''K$. So $S = CI \cap B'X = C'$, and so the points C', C'', K are collinear. Thus, K is the intersection point of $B'B''$ and $C'C''$ which implies that K is the center of the homothety mapping $A'B'C'$ to $A''B''C''$, and it belongs to ω, as claimed. This completes the proof.

Solution 2. Define the points T, A', B', and C' in the same way as in the previous solution. Let X, Y, and Z be the symmetric images of T about the lines BC, CA, and AB, respectively. Note that the projections of T on these lines form the Simson line of T with respect to ABC, therefore the points X, Y, Z are collinear. Moreover, we have $X \in B'C'$, $Y \in C'A'$, $Z \in A'B'$.

Denote $\alpha = \angle(\ell, TC) = \angle(BT, BC)$. Using the symmetry in the lines AC and BC, we get

$$\angle(BC, BX) = \angle(BT, BC) = \alpha$$

and

$$\angle(XC, XC') = \angle(\ell, TC) = \angle(YC, YC') = \alpha.$$

Since $\angle(XC, XC') = \angle(YC, YC')$, the points X, Y, C, C' lie on some circle ω_c. Define the circles ω_a and ω_b analogously. Let ω' be the circumcircle of triangle $A'B'C'$. Now, applying Miquel's theorem to the four lines $A'B'$, $A'C'$, $B'C'$ and XY, we obtain that the circles ω', ω_a, ω_b, ω_c intersect at some point K. We will show that K lies on ω, and that the tangent lines to ω and ω' at this point coincide; this implies the problem statement.

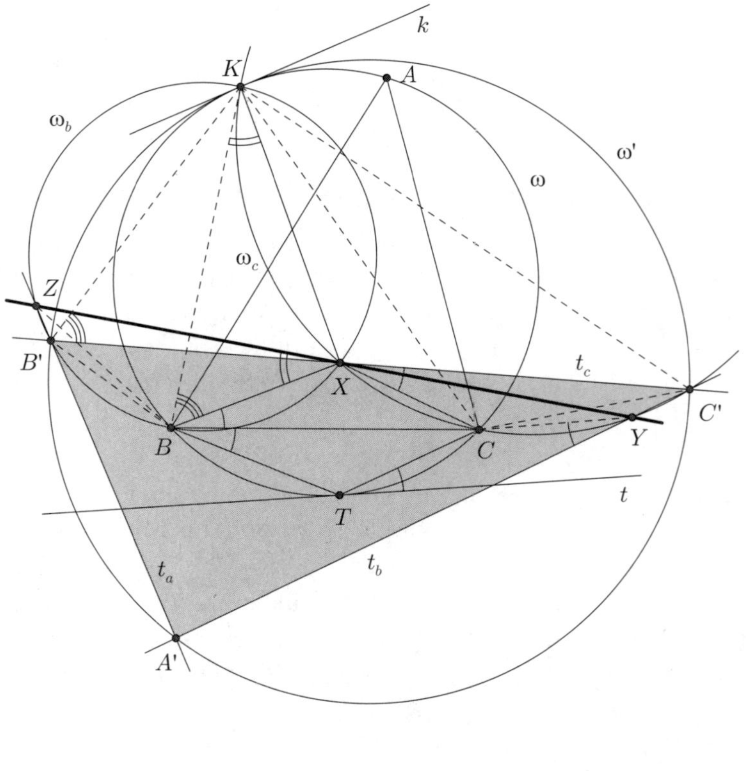

Due to symmetry, we have $XB = TB = ZB$, so the point B is the midpoint of one of the arcs XZ of circle ω_b. Therefore $\angle(KB, KX) = \angle(XZ, XB)$. Analogously, $\angle(KX, KC) = \angle(XC, XY)$. Adding these equalities and using the symmetry in the line BC we get

$$\angle(KB, KC) = \angle(XZ, XB) + \angle(XC, XZ) = \angle(XC, XB) = \angle(TB, TC).$$

Therefore, K lies on ω.

Next, let k be the tangent line to ω at K. We have

$$\begin{aligned}
\angle(k, KC') &= \angle(k, KC) + \angle(KC, KC') \\
&= \angle(KB, BC) + \angle(XC, XC') \\
&= (\angle(KB, BX) - \angle(BC, BX)) + \alpha \\
&= \angle(KB', B'X) - \alpha + \alpha \\
&= \angle(KB', B'C'),
\end{aligned}$$

which means exactly that k is tangent to ω'.

Comment. There exist various solutions combining the ideas from the two solutions presented above. For instance, one may define the point X as the reflection of T with respect to the line BC, and then introduce the point K as the second intersection point of the circumcircles of triangles $BB'X$ and $CC'X$. Using the fact that BB' and CC' are the bisectors of $\angle(A'B', B'C')$ and $\angle(A'C', B'C')$ one can show successively that $K \in \omega$, $K \in \omega'$, and that the tangents to ω and ω' at K coincide.

Solution 3. There is also a more straightforward solution using Casey's theorem found by one of the contestants from Iran.

Define the points T, A', B', and C' in the same way as in the previous solution. Denote the powers of points A', B', C' with respect to ω by p_a, p_b, p_c, respectively. By Casey's theorem, applied for A', B', C' and ω it suffices to prove that there is a choice of signs so that the following identity is true:

$$\pm B'C' \cdot \sqrt{p_a} \pm C'A' \cdot \sqrt{p_b} \pm A'B' \cdot \sqrt{p_c} = 0.$$

The plan becomes to compute $B'C'\sqrt{p_a}$. In doing so, we omit constants that are the same for the other two terms. Let d_a, d_b, d_c be the distances from vertices A, B, C to ℓ respectively. By considering the angles of triangle $A'B'C'$ and the fact that I, the incenter of triangle $A'B'C'$, lies on ω (fact which we note as in the first solution), we have

$$A'A = \frac{d_a}{\cos A} \quad \text{and} \quad A'I \sim \frac{1}{\sin \frac{A'}{2}} \sim \frac{1}{\cos A},$$

where the latter equality holds because $\angle A' = 180° - 2\angle A$, as we have seen before (either in Problem 45 (b) or in the previous proofs). Hence

$$\sqrt{p_a} \sim \frac{\sqrt{d_a}}{\cos A} \sim \frac{AT}{\cos A}.$$

Also, we have $B'C' \sim \sin A' \sim \sin 2A$; therefore

$$B'C' \cdot \sqrt{p_a} \sim AP \cdot \sin A \sim AP \cdot BC,$$

and so the assertion

$$\pm B'C' \cdot \sqrt{p_a} \pm C'A' \cdot \sqrt{p_b} \pm A'B' \cdot \sqrt{p_c} = 0$$

follows from Ptolemy's theorem.

Solution 4. During the contest, a lot of students also tried to use inversion with respect to the tangency point T but no one was able to come up with a complete solution. We therefore challenged ourselves to do it this way. What we came up with was the following generalization.

Claim. Let ABC be a triangle and let P be a point P in its plane. A line passing through P intersects the circumcircles of triangles PBC, PCA, PAB again at P_a, P_b, P_c, respectively. Let ℓ_a, ℓ_b, ℓ_c be the tangents at P_a, P_b, P_c with respect to circles (PBC), (PCA), (PAB), respectively. Then, the circumcircle of the triangle determined by the lines ℓ_a, ℓ_b, ℓ_c is tangent to the circumcircle of triangle ABC.

Proof of the Claim. Consider an inversion with center P and arbitrary power; label inverse points with primes, as usual. Also, we shall write (XYZ) for the circumcircle of triangle XYZ. Now, the given line ℓ through P is fixed and circles (PBC), (PCA) and (PAB) go to the sidelines $B'C'$, $C'A'$, $A'B'$ of triangle $A'B'C'$. ℓ cuts $B'C'$, $C'A'$, $A'B'$ at $P_a{}'$, $P_b{}'$, $P_c{}'$. Thus, tangents ℓ_a, ℓ_b, ℓ_c of (PBC), (PCA), (PAB) through P_a, P_b, P_c go to the circles $\omega_a, \omega_b, \omega_c$ passing through P and tangent to $B'C'$, $C'A'$, $A'B'$ through $P_a{}'$, $P_b{}'$, $P_c{}'$. Pairwise circles $\omega_a, \omega_b, \omega_c$ meet at D', E', F', the inverses of the vertices of $\triangle(\ell_a, \ell_b, \ell_c) \equiv \triangle DEF$. Hence, in order to show that the circumcircle of triangle DEF is tangent to the circumcircle of ABC, we shall show that $(D'E'F')$ is tangent to $(A'B'C')$.

For convenience, drop the primes from the new figure. Let M be the Miquel point of $\triangle ABC \cup \ell$. Henceforth, we will use oriented angles modulo $180°$. From the tangencies of ω_b, CA and ω_c, AB we get $\angle AP_bP_c = \angle PDP_b$ and $\angle AP_cP_b = \angle PDP_c$; therefore $D \in (AP_bP_c)$. Likewise, we have $E \in (BP_cP_a)$ and $F \in (CP_aP_b)$.

Now, the rest is just simple angle chasing using inscribed angles spanning the same arcs. More precisely:

$$\begin{aligned}\angle EFD &= \angle PFD + \angle PFE \\ &= \angle P_cP_bD + \angle P_cP_aE \\ &= \angle DMP_c + \angle EMP_c,\end{aligned}$$

or equivalently $\angle EFD = \angle EMD$, i.e., $M \in (DEF)$.

Furthermore,

$$\angle MFE = \angle EFP_a + \angle MFP_a = \angle EP_aB + \angle MCB = \angle EMB + \angle MAB,$$

so

$$90^\circ - \angle MFE + \angle EMB = 90^\circ - \angle MAB,$$

which means that the circumcenters of triangles MEF and MBA are collinear with M. This, together with the above angle calculation, yields that the circumcircles of triangles ABC and DEF are tangent at M. This completes the proof of the Claim and of the problem.

■

Problem 57. (Wilhelm Fuhrmann) Let M be a point inside triangle ABC with circumcircle (O). Let A_1, B_1, C_1 be the intersections of AM, BM, CM with (O) which are different from the vertices of the triangle. Furthermore, let A_2, B_2, C_2 be the reflections of A_1, B_1, C_1 across the sidelines BC, CA, and AB, respectively. Prove that the triangles $A_1B_1C_1$ and $A_2B_2C_2$ are similar.

Solution. The key fact is given by the following preliminary result, which can be seen as a good contest problem by itself.

Claim. If H is the orthocenter of triangle ABC, then $HB_2A_2C_2$ is cyclic.

Notice that the reflection of H into any one of the sides of triangle ABC lies on the circumcircle of ABC, meaning that AC_2HB is cyclic and similarly for the rest. Thus, inversion around H which maps point X to X' gives that A', B', and C_2' are collinear and similarly for the others. From the inversions, we have that $\frac{A'C_2'}{AC_2} = \frac{HC_2'}{HA}$ and $\frac{B'C_2'}{BC_2} = \frac{HC_2'}{HB}$. Dividing yields that $\frac{A'C_2'}{B'C_2'} \cdot \frac{BC_2}{AC_2} = \frac{HB}{HA}$. Doing the same for the rest and multiplying gives that

$$\frac{A'C_2'}{B'C_2'} \cdot \frac{B'A_2'}{A_2'C'} \cdot \frac{C'B_2'}{A'B_2'} = \frac{AC_2}{BC_2} \cdot \frac{A_2B}{A_2C} \cdot \frac{CB_2}{AB_2} = \frac{AC_1}{BC_1} \cdot \frac{A_1B}{A_1C} \cdot \frac{CB_1}{AB_1} = 1$$

from the concurrency of AA_1, BB_1, and CC_1, so by Menalaus, C_2', B_2', and A_2' are collinear, yielding that $HC_2A_2B_2$ is cyclic, proving the claim.

Now, returning to the problem, we see that

$$\begin{aligned}\angle C_2A_2B_2 &= 180^\circ - \angle C_2HB_2\\ &= 180^\circ - (\angle AHB_2 - \angle AHC_2)\\ &= 180^\circ - (180^\circ - \angle B_2CA - \angle ABC_2)\\ &= \angle ACB_1 + \angle C_1BA\\ &= \angle C_1A_1B_1,\end{aligned}$$

and similarly for the other angles, yielding that the triangles $A_1B_1C_1$ and $A_2B_2C_2$ are indeed similar.

■

Problem 58. (China IMO TST 2013) Let P be a given point inside the triangle ABC. Suppose L, M, N are the midpoints of BC, CA, AB, respectively, and

$$PL : PM : PN = BC : CA : AB.$$

The extensions of AP, BP, CP meet again the circumcircle of ABC at points D, E, F, respectively. Prove that the circumcenters of triangles PBF, PCE, PCD, PAF, PAE, PBD are concyclic.

Solution. Let A_1, A_2, B_1, B_2, C_1, C_2 be the circumcenters of triangles PBF, PCE, PCD, PAF, PAE, PBD, respectively. It is clear that the perpendiculars to AD from A and D, the perpendiculars to BE from B and E and the perpendiculars to CF from C and F bound a hexagon whose opposite sides are parallel and whose diagonals bisect each other at the circumcenter O of triangle ABC. The vertices of this hexagon are the images of A_1, A_2, B_1, B_2, C_1, C_2 under the homothety with center P and ratio 2; hence, the segments A_1A_2, B_1B_2, C_1C_2 and OP all bisect each other at K.

With this observation in mind, it is thus sufficient to show that $A_1A_2 = B_1B_2 = C_1C_2$. Using the median formula for PK in triangle PA_1A_2,

$$PK^2 = \frac{PA_1{}^2 + PA_2{}^2}{2} - \frac{A_1A_2{}^2}{4}.$$

However, the circumradii PA_1 and PA_2 satisfy

$$\frac{PB}{PA_1} = \frac{PC}{PA_2} = 2\sin A = \frac{BC}{R}.$$

Hence, it follows that

$$\begin{aligned}PK^2 &= \frac{R^2}{2BC^2}(PB^2 + PC^2) - \frac{A_1A_2{}^2}{4}\\ &= \frac{R^2}{2BC^2}\left(2PL^2 + \frac{BC^2}{2}\right) - \frac{A_1A_2{}^2}{4},\end{aligned}$$

which yields

$$A_1A_2^2 = 4R^2\left(\frac{PL^2}{BC^2}+\frac{1}{4}\right) - 4PK^2.$$

Analogously, we also get that

$$B_1B_2^2 = 4R^2\left(\frac{PM^2}{CA^2}+\frac{1}{4}\right) - 4PK^2$$

and

$$C_1C_2^2 = 4R^2\left(\frac{PN^2}{AB^2}+\frac{1}{4}\right) - 4PK^2,$$

which all combined with $\frac{PL}{BC} = \frac{PM}{CA} = \frac{PN}{AB}$ give $A_1A_2 = B_1B_2 = C_1C_2$, as desired. This completes the proof.

■

Problem 59. (Luis Gonzalez, Mathematical Reflections) Let ABC be an acute-angled triangle and let τ be the inradius of its orthic triangle. Prove that

$$r \geq \sqrt{R\tau},$$

where r and R are the inradius and circumradius of triangle ABC, respectively.

Solution. We begin with a classical geometric inequality.

Lemma. In any triangle ABC, we have

$$\cos A + \cos B + \cos C \leq \frac{3}{2},$$

with equality if and only if ABC is equilateral.

There are many possible proofs, from using identities such as

$$\cos A + \cos B + \cos C = 1 + \frac{r}{R}$$

to employing algebraic tools such as Jensen's Inequality. We include a different short proof below for the reader's convenience.

Proof. The inequality may be written as

$$4\sin\frac{A}{2}\cos\frac{B-C}{2} \leq 1 + 4\sin^2\frac{A}{2},$$

clearly true since the RHS is, by the AM-GM inequality, at least $4\sin\frac{A}{2}$, and the LHS is at most $4\sin\frac{A}{2}$, with equality in both cases if and only if $\sin\frac{A}{2} = \frac{1}{2}$ and $\cos\frac{B-C}{2} = 1$, i.e., if and only if ABC is equilateral. The Claim follows.

Returning to the problem, denote by D, E, F the respective feet of the altitudes from A, B, C. It is well-known (and easy to show) that $DE = c\cos C$, $EF = a\cos A$ and $FD = b\cos B$, and moreover $\angle DEF = 180° - 2\angle B$, $\angle EFD = 180° - 2\angle C$ and $\angle FDE = 180° - 2\angle A$. The perimeter of the orthic triangle is then

$$a\cos A + b\cos B + c\cos C = \frac{2a^2b^2 + 2b^2c^2 + 2c^2a^2 - a^4 - b^4 - c^4}{2abc} = \frac{2[ABC]}{R},$$

where S, R are the area and circumradius of ABC, and both Heron's formula and the well known relation $S = \frac{abc}{4R}$ have been used. The area of the orthic triangle is easily found to be $2[ABC]\cos A\cos B\cos C$, or

$$\tau = 2R\cos A\cos B\cos C,$$

while it is well known (or easily found, for example comparing different expressions for the area of ABC) that $r = 4R\sin\frac{A}{2}\sin\frac{B}{2}\sin\frac{C}{2}$. Using that $2\sin^2 x = 1 - \cos(2x)$, and assuming (without loss of generality) that $A \le B \le C$, or $2\cos A \ge 1$, we may rewrite the proposed inequality as

$$(1 - \cos A)(1 - \cos B - \cos C) \ge \cos B\cos C(2\cos A - 1).$$

Now, $1 - \cos B - \cos C \ge \frac{2\cos A - 1}{2}$ by the Claim, or it suffices to show that

$$1 \ge \cos A + 2\cos B\cos C = \cos(B - C),$$

clearly true and with equality if and only if $B = C$. Now, when $\cos B\cos C = \frac{1-\cos A}{2}$, the proposed inequality is equivalent to the Claim. The conclusion follows, and equality holds if and only if ABC is equilateral, in which case $R = 2r = 4\tau$.

■

Problem 60. (Christopher Bradley, IMO Shortlist 2007) Let ABC be a fixed triangle, and let A_1, B_1, C_1 be the midpoints of sides BC, CA, AB, respectively. Let P be a variable point on the circumcircle. Let lines PA_1, PB_1, PC_1 meet the circumcircle again at A', B', C', respectively. Assume that the points A, B, C, A', B', C' are distinct, and lines AA', BB', CC' form a triangle. Prove that the area of this triangle does not depend on P.

Solution. Assume triangle ABC and P are chosen like in the below diagram. Denote by X, Y, Z the intersections $BB' \cap CC'$, $CC' \cap AA'$ and $AA' \cap BB'$, respectively. By Pascal's theorem on the cyclic hexagon $ABB'PC'C$, the points B_1, X, C_1 are collinear. Similarly, the points C_1, Y, A_1 are collinear and so are A_1, Z, B_1. The lines AY, XB, B_1A_1 are concurrent at Z, and so triangles AXB_1 and YBA_1 are perspective. Desargues' theorem in this case tells us that the intersections

$$AX \cap YB,\ XB_1 \cap BA_1 = \infty, \text{ and } AB_1 \cap YA = \infty$$

are collinear, or in other words, $AX \| BY$. Analogously, we get that $BY \| CZ$, which enables us to write that $AX \| BY \| CZ$.

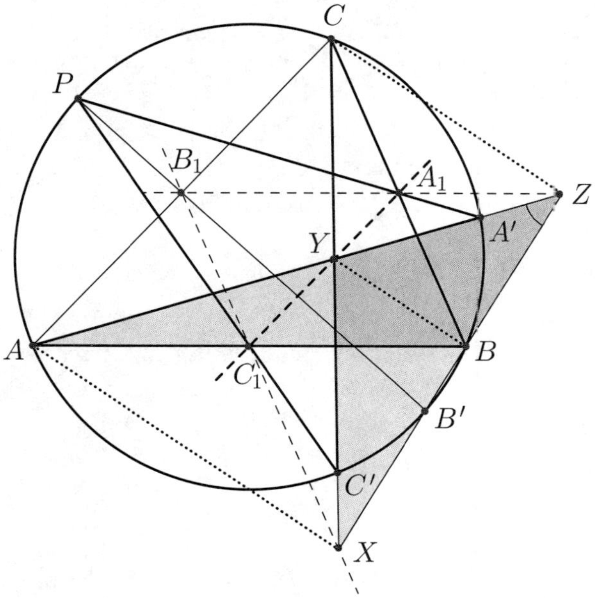

This, combined with the collinearity of Y, C_1 and A_1 enables us to write

$$\begin{aligned} [XYZ] &= [XYB] + [YZB] \\ &= [YAB] + [YBC] \\ &= 2[YC_1B] + 2[YBA_1] \\ &= 2[BC_1A_1] \\ &= \frac{1}{2}[ABC]. \end{aligned}$$

This is a value that is independent of the position of P, so the proof is complete. ■

Comment. The configuration contains numerous other interesting properties. For instance, consider the parallels through vertices A, B, C to the sides BC, CA, AB and let MNQ be the triangle bounded by these lines. Let α, β, γ be the reflections of lines NQ, QM, MN across AX, BY, CZ, respectively. Then, we claim that

(a) α, β, γ concur at some point P', which lies on the circumcircle of ABC;
(b) $PP' \| AX \| BY \| CZ$, where P' is the concurrence point from part (a).

Similar properties hold when instead of taking parallels, we consider α, β, γ to be the sidelines of the excentral triangle $I_aI_bI_c$. We leave the reader with the task of finding the right statement(s) and the corresponding proofs.

Problem 61. (Antreas Hatzipolakis, Hyacinthos Newsgroup) Let P be a point in the plane of a triangle ABC. The lines AP, BP, CP intersect the lines BC, CA, AB at the points A', B', C'. Let Q be the isogonal conjugate of the point P with respect to triangle ABC. Then, the reflections of the lines AQ, BQ, CQ in the lines $B'C'$, $C'A'$, $A'B'$ are concurrent.

Solution. We begin with a result by Jean-Pierre Ehrmann.

Lemma. Let P and Q be two isogonal conjugates with respect to a triangle ABC. Then, the reflection of the line AP with respect to the internal angle bisector of $\angle BPC$ and the reflection of the line AQ with respect to internal angle bisector of $\angle BQC$ are symmetric to each other with respect to the sideline BC.

Proof. Let X', Y', Z' be the reflections of the point P across the lines BC, CA, AB, respectively, and let U' be the reflection of the point Q in the line BC. Then, the point Q is, in turn, the reflection of U' in the line BC. And since the points Q and X' are the reflections of U' and P in the line BC, the line QX' is the reflection of the line $U'P$ in the line BC. This means that the lines PU' and QX' are symmetric to each other with respect to the line BC.

However, by Lemma 2 of Problem 18, the point Q is the circumcenter of triangle $X'Y'Z'$, thus the center of the circle through the points X', Y', Z'. Hence, we get that $\angle QX'Z' = 90° - \angle Z'Y'X'$. In other words, $\angle(QX', Z'X') = 90° - \angle(Y'Z', X'Y')$. Furthermore, Theorem B also tells us that the lines AQ, BQ, CQ are the perpendicular bisectors of the segments $Y'Z'$, $Z'X'$, $X'Y'$, thus $\angle(Y'Z', AQ) = 90°$, $\angle(BQ, Z'X') = 90°$ and $\angle(X'Y', CQ) = 90°$. Hence, we immediately get that $\angle(BQ, QX') = \angle(CQ, AQ)$. Thus, the line QX' is the reflection of the line AQ with respect to the angle $\angle BQC$. Similarly, the line PU' is the isogonal conjugate of AP with respect to $\angle BPC$. And since we know that the lines PU' and QX' are symmetric to each other with respect to BC, the conclusion follows. This proves the Lemma.

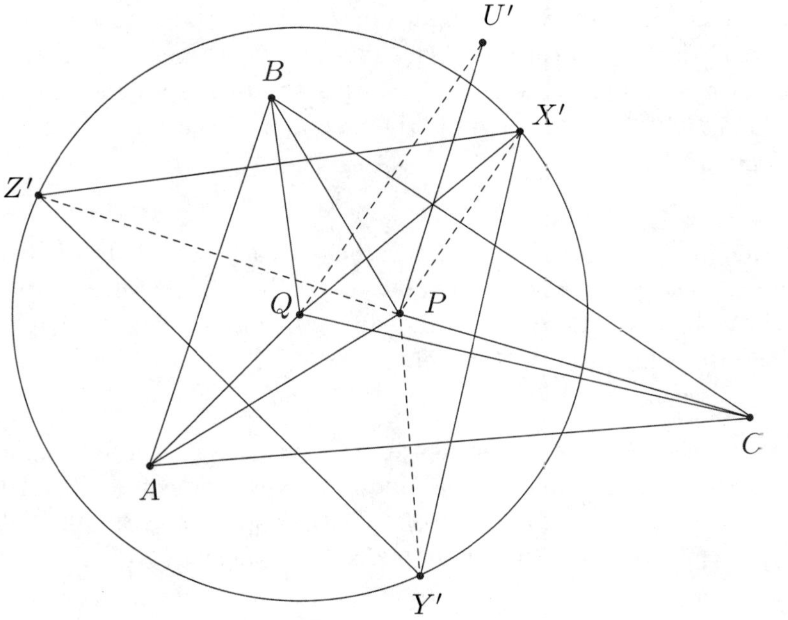

Returning to the problem, let A_1, B_1, C_1, P_1 be the isogonal conjugates of the points A, B, C, P with respect to triangle $A'B'C'$, respectively.

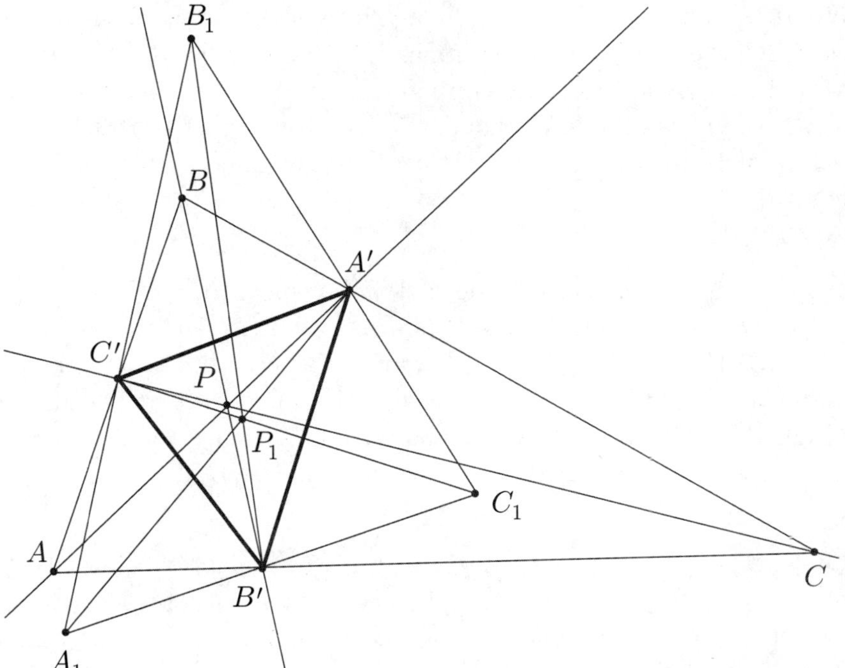

Since P_1 is the isogonal conjugate of P with respect to triangle $A'B'C'$, the line $A'P_1$ is the reflection of the line $A'P$ with respect to the internal angle bisector of angle $\angle C'A'B'$. Since A_1 is the isogonal conjugate of A with respect to $A'B'C'$, the line $A'A_1$ is the isogonal of the line $A'A$ with respect to the angle $C'A'B'$.

But since the lines $A'P$ and $A'A$ coincide, their isogonals with respect to the angle $C'A'B'$ must also coincide; i.e., the lines $A'P_1$ and $A'A_1$ coincide. Hence, the points A', A_1 ,P_1 are collinear, and similarly we get that the points B', B_1, P_1 are collinear, and the points C', C_1, P_1 are collinear.

Since B_1 is the isogonal conjugate of B with respect to triangle $A'B'C'$, the line $A'B_1$ is the isogonal of the line $A'B$ with respect to $\angle C'A'B'$. Also, since C_1 is the isogonal conjugate of C with respect to triangle $A'B'C'$, the line $A'C_1$ is the isogonal of the line $A'C$ with respect to $\angle C'A'B'$.

But again, since the lines $A'B$ and $A'C$ coincide, their isogonal with respect to $\angle C'A'B'$ need to coincide, so $A'B_1$ and $A'C_1$ coincide. Thus, we have that the points A', B_1, C_1 are collinear, and similarly, the points B', C_1, A_1 are collinear, and the points C', A_1, B_1 are collinear.

Now, the point Q is the isogonal conjugate of the point P with respect to triangle ABC, the line AQ is the isogonal of the line AP w.r.t. the angle CAB. However, A and A_1 are isogonal conjugates with respect to $A'B'C'$, thus the Lemma above yields that the isogonal of $A'A$ with respect to $\angle B'AC'$ and the isogonal of $A'A_1$ with respect to $\angle B'A_1C'$ are symmetric to each other with respect to $B'C'$.

Moreover, the isogonal of the line $A'A$ with respect to $B'AC'$ is the isogonal of the line AP with respect to CAB (since the line $A'A$ is the line $A'P$, and the angle $\angle B'AC'$ is the angle $\angle CAB$), and this is the line AQ. Furthermore, the isogonal of the line $A'A_1$ with respect to the angle $\angle B'A_1C'$ is the isogonal of the line A_1P_1 with respect to the angle $C_1A_1B_1$ (since the line $A'A_1$ is the line A_1P_1 and the angle $\angle B'A_1C'$ is the angle $\angle C_1A_1B_1$).

Thus, we got that the line AQ and the isogonal of the line A_1P_1 with respect to $\angle C_1A_1B_1$ are symmetric to each other with respect to the line $B'C'$. In other words, the reflection of AQ in the line $B'C'$ is the isogonal of the line A_1P_1 with respect to $\angle C_1A_1B_1$.

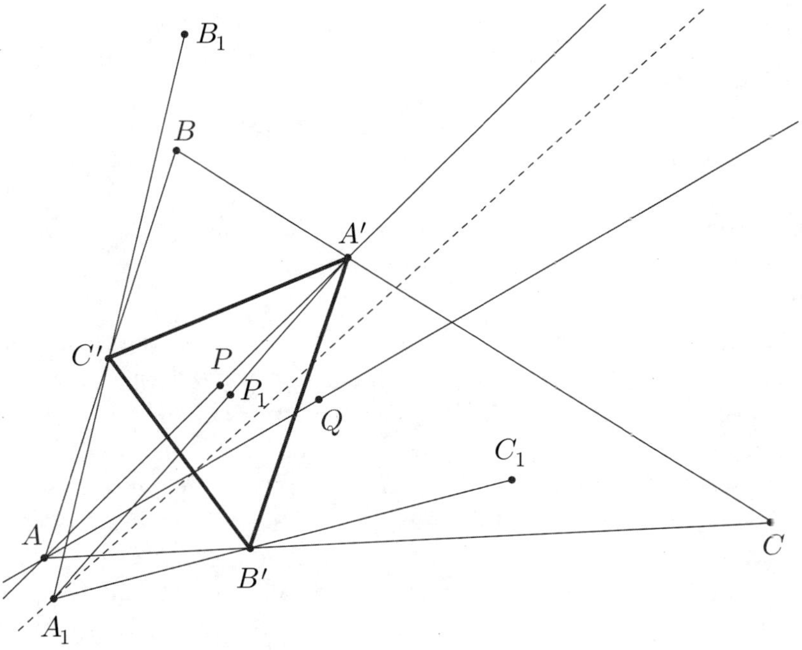

Similarly, we can say the analogues about the reflections of the lines BQ and CQ across $C'A'$ and $A'B'$; thus, we conclude that the reflections of the lines AQ, BQ, CQ in the lines $B'C'$, $C'A'$, $A'B'$ are concurrent, as these lines are the isogonals of the concurrent lines A_1P_1, B_1P_1, C_1P_1 with respect to the angles of triangle $A_1B_1C_1$, so they need to meet at the isogonal conjugate of the point P_1 with respect to triangle $A_1B_1C_1$. This completes the proof.

■

Problem 62. (Darij Grinberg and Cosmin Pohoata, Mathematical Reflections) Let triangle ABC be a triangle with circumcircle (O) and incircle (I). Let D, E, F be the tangency points of (I) with BC, CA, AB, respectively. The line EF intersects the circumcircle (O) at X_1, X_2. Similarly, we define Y_1, Y_2 and Z_1, Z_2. Prove that the radical center of the circumcircles of triangles DX_1X_2, EY_1Y_2, FZ_1Z_2 is the orthocenter of triangle DEF.

Solution 1. Let X_1X_2 intersect BC at X and let the midpoint of BC be M. Let the midpoints of AB and AC be P and N respectively. Let the centers of (EY_1Y_2) and (FZ_1Z_2) be O_1 and O_2 respectively. Let the parallel to BI through O intersect IE at P and the parallel to IC through O meet IF at Q.

We claim that PQ is parallel to EF. Indeed, let the feet of the perpendiculars from E, O, and P to BI be E', O', and P', respectively. It is known that if H is the orthocenter of triangle ABC, then O and H are isogonal conjugates with respect to ABC. Thus, since $IE \parallel BH$, we have that $\angle EIE' = \angle OBO'$. Furthermore, $\angle OO'B = \angle EE'I$, so $\triangle EE'I \sim \triangle OO'B$. This means that

$$\frac{PI}{IE} = \frac{PP'}{EE'} = \frac{OO'}{EE'} = \frac{OB}{EI} = \frac{R}{r}$$

where R is the circumradius of $\triangle ABC$ and r is its inradius. Similarly, $\frac{IQ}{IF} = \frac{R}{r} = \frac{PI}{IE}$, so $PQ \parallel FE$, which proves the claim.

Returning to the problem, we now look at the complete quadrilateral $XBCEAF$. The quadruple (X, B, D, C) is a harmonic division. This implies that the pencil (XB, XD, XM, XC) is harmonic, since M is the midpoint of BC, and hence $XX_1 \cdot XX_2 = XB \cdot XC = XD \cdot XM$; thus X_1X_2MD is cyclic. Similarly, the circumcircle of $\triangle EY_2Y_1$ passes through N and that of $\triangle FZ_1Z_2$ passes through P. This means that O_1 is the intersection of the perpendicular bisector of EN and the perpendicular bisector of Y_1Y_2. Notice that the latter is the perpendicular to DF through O. Since $IB \perp DF$, we have that it is the parallel to IB through O. Therefore, O_1 is the intersection of the perpendicular through the midpoint of EN to AC and the parallel to BI through O. It follows that O_1 is the midpoint of OP. Similarly, O_2 is the midpoint of OP, so $O_1O_2 \parallel PQ \parallel EF$. This means that the radical axis of (EY_1Y_2) and (FZ_1Z_2) is perpendicular to O_1O_2, so it is perpendicular to EF. Now, consider the circles $(Y_1Z_2Y_1Z_1)$, (Y_1Y_2E), and (Z_1Z_2F). The radical axis of the latter two circles must pass through D, and since this is the radical center, it therefore follows that the radical axis of (Y_1Y_2E) and (Z_1Z_2F) must be the perpendicular through D to EF. Doing the same for the other radical axes, we see that they must concur at the orthocenter of triangle DEF. This completes the first proof.

Solution 2. Let S be the intersection of EF with the sideline BC, and M is the midpoint of BC. We have the division $(BDCS)$ is harmonic, so that $SD \cdot SM = SB \cdot SC = SX_1 \cdot SX_2$. Thus, M lies on the circumcircle of triangle DX_1X_2.

Let us call P, Q the midpoints of arcs BC and AB of the circle $\rho(O)$ (not containing A, C), respectively. Let N the midpoint of the arc BC containing A of $\rho(O)$, O_1 the midpoint of the segment IN. We have OO_1 is parallel to IP, which means OO_1 is perpendicular to X_1X_2, and so it's also the perpendicular bisector of X_1X_2. On the other hand, we have the quadrilateral $IDMN$ is a trapezoid, so O_1 lies on the perpendicular bisector of DM. Now, we have that the quadrilateral DMX_2X_1 is cyclic and O_1 is the intersection of the perpendicular bisectors of the segments DM and X_1X_2, so O_1 is the center of the circumcircle of triangle DX_1X_2.

We can easily see that OO_1 is parallel to IP and $OO_1 = \frac{1}{2} \cdot IP$. Similarly, OO_3 is parallel to IQ and $OO_3 = \frac{1}{2} \cdot IQ$, where O_3 is the center of the circumcircle of triangle FZ_1Z_2. Hence, we obtain O_1O_3 is parallel to PQ. Now, let's take a look at the circumcircles of triangles DX_1X_2 and FZ_1Z_2, we have the radical axis of the circles is the line d_2, which passes through E and perpendicular to O_1O_3, i.e., $d_2 \perp PQ$. It follows d_2 is the E-altitude of triangle DEF. Similarly, we have the radical axis d_3 of the circumcircles of triangles DX_1X_2 and EY_1Y_2 is the F-altitude of triangle DEF. Hence, we deduce the radical center of the circles DX_1X_2, EY_1Y_2, and FZ_1Z_2 is the orthocenter of triangle DEF, which we know lies on OI. This completes the second proof.

■

Problem 63. (Cosmin Pohoata, Mathematical Reflections) Let M be an arbitrary point on the circumcircle of triangle ABC and let the tangents from this point to the incircle of the triangle meet the sideline BC at X_1 and X_2. Prove that the second intersection of the circumcircle of triangle MX_1X_2 with the circumcircle of ABC coincides with the tangency point of the circumcircle with the A-mixtilinear incircle.

Solution 1. We will use inversion with respect to the incenter. We isolate a result that proves to be very useful when dealing with complicated configurations involving the mixtilinear incircles.

Lemma. Let ABC be a triangle with incircle $\mathcal{I}(I, r)$ and consider the inversion Ψ with center I and power r^2. Let A_1, B_1, C_1 be the tangency points of the mixtilinear incircles $\mathcal{K}_a$, $\mathcal{K}_b$, $\mathcal{K}_c$ with the circumcircle $\mathcal{O}$. Then,

$$\Psi(A_1) = A_1', \ \Psi(B_1) = B_1', \ \Psi(C_1) = C_1'$$

are the antipodes of the points $\Psi(A)$, $\Psi(B)$, $\Psi(C)$ with respect to the circumcircle of triangle $\Psi(A)\Psi(B)\Psi(C)$.

Proof of Lemma. Let D, E, F be the tangency points of the incircle (I) with the sides BC, CA, AB, respectively, and let A', B', C' be the midpoints of sides EF, FD, DE of triangle DEF. Recall that the images of vertices A, B, C under Ψ are the points A', B', C', respectively. Or in other words, $\Psi(A) = A'$, $\Psi(B) = B'$, $\Psi(C) = C'$. Thus, showing that A_1' is the antipode of A' in the circumcircle of triangle $A'B'C'$ is equivalent to checking that

$$\begin{aligned}
90^\circ &= \angle A'B'A_1' \\
&= \angle IB'A' + \angle IB'A_1' \\
&= \angle IAB + \angle IA_1B \\
&= \frac{1}{2}\angle A + \angle IA_1B.
\end{aligned}$$

But recall from Problem 16 that the line A_1I is the internal angle bisector of $\angle BA_1C$; hence

$$\angle IA_1B = \frac{1}{2}\angle BA_1C = \frac{1}{2}(180^\circ - \angle A) = 90^\circ - \frac{1}{2}\angle A.$$

This proves indeed that A_1' is the antipode of A' in the circumcircle of triangle $A'B'C'$, and similarly for B_1' and C_1'.

□

Returning to the problem, assume without loss of generality that M lies on the same side of line BC as the vertex A. Let (I) be the common incircle of triangles ABC and MX_1X_2, let r be its radius, and let D, E, F, Y_1, Y_2 be the tangency points of (I) with the sides BC, CA, AB, MX_1, MX_2, respectively.

Consider the inversion Ψ with center I and power r^2, which takes the vertices A, B, C, M, X_1, X_2 to the midpoints A', B', C', M', X_1', X_2' of the sides EF, FD, DE, Y_2Y_1, Y_1D, DY_2 of the intouch triangles DEF and DY_2Y_1. The circumcircles (O), (P) of triangles ABC, and MX_1X_2 become the circumcircles (O'), (P') of triangles $A'B'C'$, $M'X_1'X_2'$, which, because they coincide with the nine-point circles of two triangles of the same circumcircle, are congruent and have the common radius $\frac{r}{2}$.

Let Γ_a, Γ_b, Γ_c be the circles passing through I that are the images of sides BC, CA, AB under Ψ. These are tangent to BC, CA, AB at the points D, E, F, respectively, and so they all have inradii equal to the inradius r of ABC. They pairwise intersect at the midpoints of A', B', C' of sides EF, FD, DE of the contact triangle, since these are the images of A, B, C under Ψ.

Let A_0, B_0, C_0, O_1, O_2 be the centers of the congruent circles Γ_a, Γ_b, Γ_c, Ω_1, Ω_2, respectively. Since Γ_a is the reflection of (O_1) in $B'C'$ and since O_1 and I are isogonal conjugated with respect to triangle $A'B'C'$, the quadrilateral $B'Z'C'I$ is a parallelogram, and therefore, its diagonals $B'C'$, IZ' cut each other in half at the midpoint of segment $B'C'$. It now follows that the quadrilateral $IA_0Z'O'$ is also a parallelogram, and thus, $IA_0 = O'Z' = \frac{r}{2}$. Now since $A_0I = A_0X_1' = O_1I = O_1X_1' = \frac{r}{2}$, the quadrilateral $A_0IO_1X_1'$ is a rhombus, and since the segments O_1X_1', IA_0, $O'Z'$ are parallel and congruent, the quadrilateral $O_1X_1'Z'O'$ is a parallelogram. In conclusion, the triangles $PX_1'Z'$, and $M'O_1O'$ are congruent and $P'Z' = M'O' = \frac{r}{2}$. Hence, Z' lies on the circle (P'), which means that the second intersection of the circumcircle (P) of triangle MX_1X_2 with the circumcircle (O) of ABC (different from M) coincides with the tangency point Z of the mixtilinear incircle in angle A with the circumcircle (O).

■

Comment. As a matter of fact, the following more general result holds.

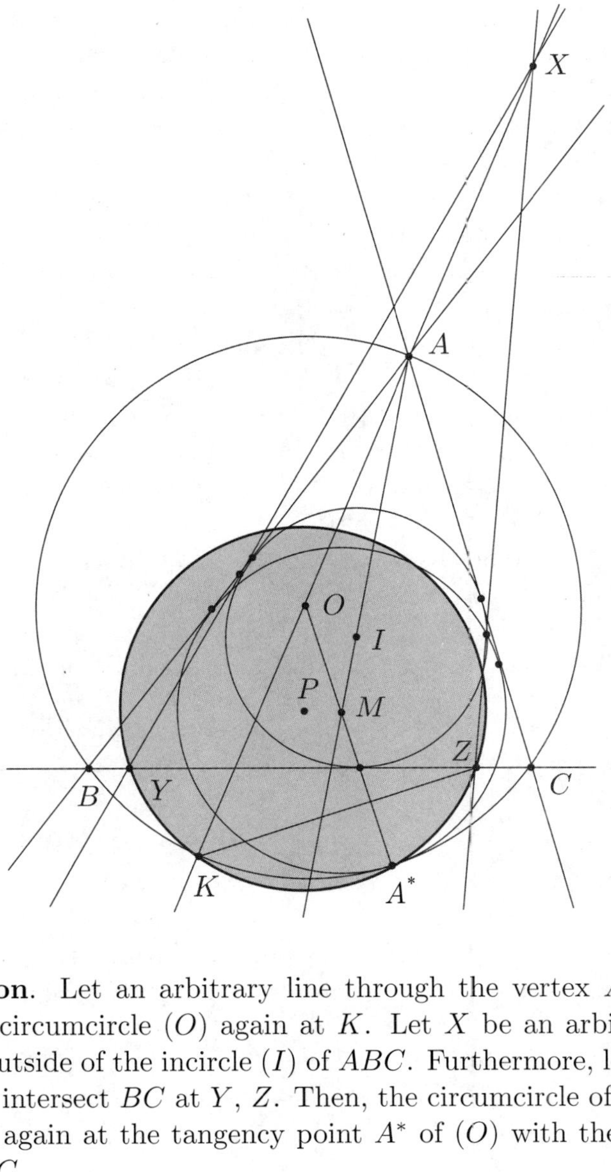

Generalization. Let an arbitrary line through the vertex A of a triangle ABC cut the circumcircle (O) again at K. Let X be an arbitrary point on the line AK, outside of the incircle (I) of ABC. Furthermore, let the tangents from X to (I) intersect BC at Y, Z. Then, the circumcircle of triangle KYZ intersects (O) again at the tangency point A^* of (O) with the A-mixtilinear incircle of ABC.

The proof uses the result of Problem 63, which we just proved. We however need to be very careful about the reduction.

Proof. Let (I) be an arbitrary fixed circle with fixed parallel tangent lines a and a', and let k be an arbitrary fixed line, which may or may not intersect the circle. Let X be arbitrary point on k and let the tangents to (I) from X cut a at Y, Z so that (I) is incircle of triangle XYZ. [This is not completely general; X has to be on the opposite side of a' from (I), but proofs for the other cases—X between a, a' outside of (I), or X on the opposite side of a from (I)—are the same, modulo incircle/excircle interchanges.]

Let S, S' be the intersections of k with a, a', respectively, and let the tangent to (I) from S' other than a' cut a at T. Let (I_x) be the X-excircle of triangle XYZ. Since $S'T$ is a fixed tangent to (I), the line s parallel to $S'T$ through S is tangent to (I_x). So the X-excircle is tangent to fixed lines a, s, which means that it belongs to a file of circles with the common similarity center S and two common tangents a, s. In any case, it is centered on the bisector i of the angle $\angle as$ formed by the lines a, s. Let M be the midpoint of II_x. This is the circumcenter of the cyclic quadrilateral IYI_xZ and its circumcircle (M) is centered on a parallel $m \parallel i$ at half the distance from I than the angle bisector i. But one point of the circle (M) is fixed—the incenter I—and therefore, the reflection $J \in (M)$ of I in the line m is also fixed. Hence, the circle (M) belongs to a pencil of circles defined by the fixed points I, J.

Let IJ cut $a \equiv YZ$ at T'. Then,$T'Y \cdot T'Z = T'I \cdot T'J = \text{const}$. When X coincides with S', Y coincides with T and Z is at infinity as $Z^* \equiv a' \cap a$. Since $T'T \cdot T'Z^* = \text{const}$ and $T'Z^* = \infty$, if follows that $T'T = 0$ and T', T are thus identical. Let $K \in k$ be an arbitrary fixed point (as opposed to $X \in k$ an arbitrary movable point) and let the fixed circumcircle of KIJ cut the fixed line KT again at L. Then $TY \cdot TZ = TI \cdot TJ = TK \cdot TL$ and it follows that the quadrilateral $KYLZ$ is cyclic and moreover, its circumcircle passes through two fixed points K, L, i.e., belongs to a pencil defined by these two points. We are almost done.

Let A^* be the tangency point with (O) of the A-mixtilinear incircle of triangle ABC in the angle $\angle A$. When X coincides with A, then Y, Z coincide with B, C, the circumcircle $(KYZ) \equiv (KBC)$ coincides with the circumcircle (O) and obviously, $A^* \in (O)$. When X coincides with K, the particular case of our problem tells us that the circumcircle of triangle KYZ passes through A^*. Therefore, $K, L \equiv A^*$ are the two fixed points of the pencil of all circumcircles (KYZ), where $Y, Z \in BC$ are feet of the incircle tangents from $X \in k$. This completes the proof.

Problem 64. (Lev Emelyanov, Forum Geometricorum) Let A_1, B_1, C_1 be points on the sides BC, CA, AB of triangle ABC such that the lines AA_1, BB_1, CC_1 are concurrent. Construct three circles Γ_1, Γ_2, and Γ_3 outside the triangle such that they are tangent to the sides of ABC at A_1, B_1, and C_1, respectively, and also tangent to the circumcircle of ABC. Prove that the circle tangent externally to these three circles is also tangent to the incircle of triangle ABC.

Solution. The plan is to apply Casey's theorem for circles Γ_1, Γ_2, Γ_3 and the incircle Γ_4 of triangle ABC. We denote by t_{ij} the length of a common tangent between circles Γ_i and Γ_j.

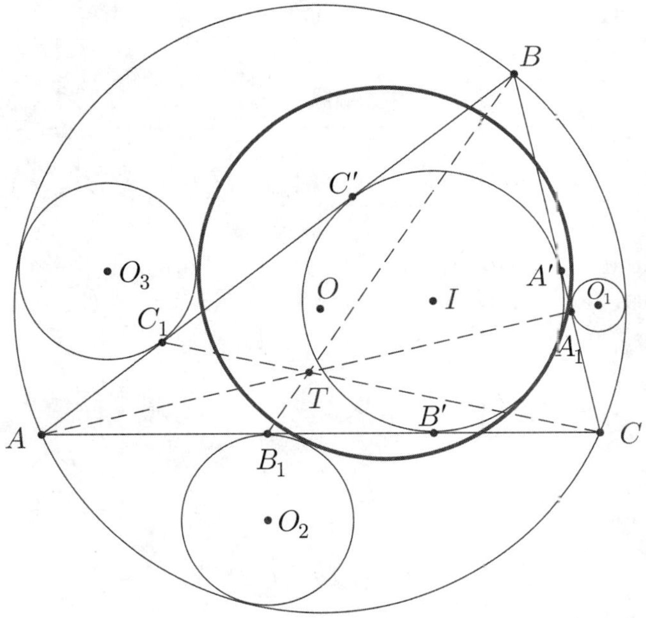

With reference to Figure 1, we show that

$$t_{12}t_{34} - t_{13}t_{42} - t_{14}t_{23} = 0,$$

where t_{12}, t_{13} and t_{23} are the lengths of the common extangents, t_{34}, t_{24} and t_{14} are the lengths of the common intangents.

Let (A) be the degenerate circle A and $t_i(A)$ be the length of the tangent from A to Γ_i. Similar notations apply to vertices B and C. Applying Casey's theorem to circles (A), (B), Γ_1 and (C), which are all tangent to the circumcircle, we have

$$t_1(A) \cdot a = c \cdot CA_1 + b \cdot A_1B.$$

From this we obtain $t_1(A)$, and similarly $t_1(B)$ and $t_1(C)$:

$$t_1(A) = \frac{c \cdot CA_1 + b \cdot A_1B}{a},$$

$$t_1(B) = \frac{a \cdot AB_1 + c \cdot B_1C}{b},$$

$$t_1(C) = \frac{b \cdot BC_1 + a \cdot C_1A}{c}.$$

Applying Casey's theorem to circles (B), (C), Γ_2 and Γ_3 we have

$$t_2(B)t_2(C) = a \cdot t_{23} + CB_1 \cdot C_1B.$$

Using the above relations, we obtain t_{23}, and similarly t_{13} and t_{12}:

$$t_{23} = \frac{a \cdot C_1A \cdot A_1B + b \cdot AB_1 \cdot BC_1 + c \cdot AC_1 \cdot CB_1}{bc},$$

$$t_{13} = \frac{b \cdot A_1B \cdot BC_1 + c \cdot BC_1 \cdot CA_1 + a \cdot BA_1 \cdot AC_1}{ca},$$

$$t_{12} = \frac{c \cdot B_1C \cdot CA_1 + a \cdot CA_1 \cdot AB_1 + b \cdot CB_1 \cdot BA_1}{ab}.$$

In the layout of Figure 1, with A', B', C' the touch points of the incircle with the sides, the lengths of the common tangents of the circles Γ_1, Γ_2, Γ_3 with the incircle are

$$t_{14} = A_1A' = -CA_1 + CA' = -CA_1 + \frac{a+b-c}{2},$$

$$t_{24} = B_1B' = -AB_1 + AB' = -AB_1 + \frac{b+c-a}{2},$$

$$t_{34} = C_1C' = BC_1 - BC' = BC_1 - \frac{c+a-b}{2}.$$

Substituting the above relations into the latter three, we get

$$t_{12}t_{34} - t_{13}t_{24} - t_{14}t_{23} = \frac{F(a,b,c)}{abc} \cdot (AB_1 \cdot BC_1 \cdot CA_1 - A_1B \cdot B_1C \cdot C_1A),$$

where

$$F(a,b,c) = 2ab + 2bc + 2ca - a^2 - b^2 - c^2.$$

However, the lines AA_1, BB_1, CC_1 are concurrent, so Ceva's theorem yields

$$AB_1 \cdot BC_1 \cdot CA_1 = A_1B \cdot B_1C \cdot C_1A.$$

In particular, this means that

$$t_{12}t_{34} = t_{13}t_{24} + t_{14}t_{23},$$

and thus the proof is completed.

■

Problem 65. (Darij Grinberg and Cosmin Pohoata, Harvard College Math. Review) Let ABC be a triangle and let P be a point in its plane. Let D, E, F be three points on the lines through P perpendicular to the lines BC, CA, and AB, respectively. Prove that if triangle DEF is equilateral and if P lies on the Euler line of ABC, then the centroid of DEF also lies on the Euler line of ABC.

Solution. Let O be the circumcenter of triangle ABC. By parallel translation, it suffices to consider the case $P = O$. Let A', B', C' be centroids of equilateral triangles outside of ABC, drawn on sides BC, CA, AB, respectively. Recall the classical

Napoleon's Theorem. Under the notations described above, triangle $A'B'C'$ is equilateral.

For sake of completeness, we include a proof below.

Proof. We use the Law of Cosines in triangle $AB'C'$ (and later in triangle ABC) to write that

$$\begin{aligned}
(B'C')^2 &= (AB')^2 + (AC')^2 - 2AB'AC' \cos \angle B'AC' \\
&= \frac{b^2}{3} + \frac{c^2}{3} - \frac{2bc}{3} \cos (A + 60^\circ) \\
&= \frac{b^2}{3} + \frac{c^2}{3} - \frac{2bc}{3} (\cos A \cos 60^\circ - \sin A \sin 60^\circ) \\
&= \frac{b^2}{3} + \frac{c^2}{3} - \frac{bc}{3} \cos A + \frac{2bc \sin A \sin 60^\circ}{3} \\
&= \frac{b^2}{3} + \frac{c^2}{3} - \frac{b^2 + c^2 - a^2}{6} + \frac{2\sqrt{3}}{3}[ABC] \\
&= \frac{a^2 + b^2 + c^2}{6} + \frac{2\sqrt{3}}{3}[ABC].
\end{aligned}$$

This expression is symmetric in a, b, c, so we can deduce that we will similarly have that

$$C'A' = A'B' = \frac{a^2 + b^2 + c^2}{6} + \frac{2\sqrt{3}}{3}[ABC]$$

too. Hence, we conclude that $O_aO_b = O_bO_c = O_cO_a$, as desired.

□

We now claim that triangles $A'B'C'$ and DEF are similar with respect to O. In particular, this means that the centroid of $A'B'C'$, the centroid of DEF, and O are collinear. Indeed, to see this, we can assume without loss of that $\angle BAC \neq 120^\circ$. Scale up or down triangle DEF with respect to O to form triangle $D'E'F'$ so that D' coincides with A'. We want to show $E' = B'$ and $F' = C'$.

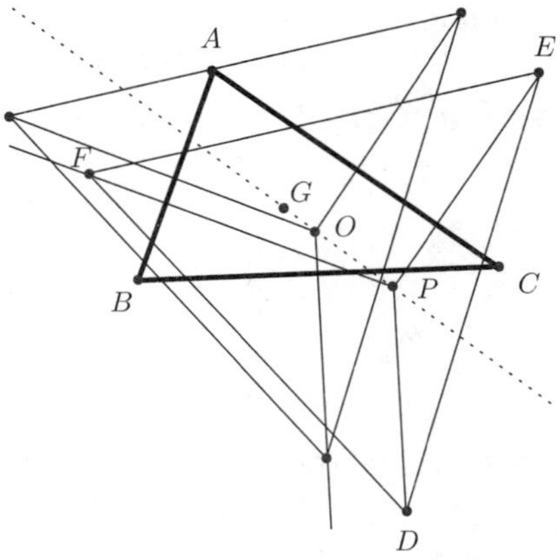

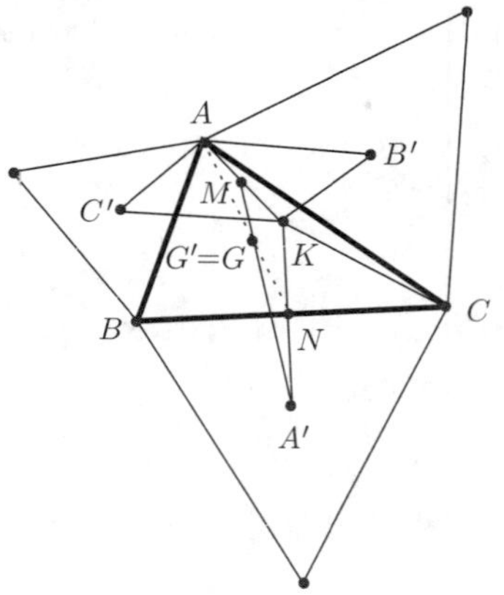

Rotate line OB' by 60° (or -60°) depending on orientation around A' and denote it by ℓ. Points B' and E' are moved on ℓ by this rotation. On the other hand, B' and E' are moved to C' and E', respectively, and these points are both on line OC'. Thus, F' and C' must be the intersection of line OC' and ℓ (note that these lines intersect at one point because $\angle BAC \neq 120^\circ$). Therefore, $C' = F'$ and $B' = E'$. It follows that triangles $A'B'C'$ and DEF are similar with respect to O.

The second claim is that the centroid of $A'B'C'$ coincides with the centroid of ABC. Indeed, let K be a point symmetric to A' with respect to line BC. Triangles $B'KC$ and ABC are similar with ratio $1 : \sqrt{3}$. In addition,

$$\frac{AC'}{AB} = \frac{1}{\sqrt{3}}.$$

Thus, $AC' = B'K$. Similarly, $AB' = C'K$, so $AC'KB'$ is a parallelogram.

Let the centroid of $AC'KB'$ be M, and the midpoint of BC be N. The centroid G' of $A'B'C'$ lies on MA' and $\frac{MG'}{G'A'} = \frac{1}{2}$. Thus,

$$\frac{A'N}{NK} \cdot \frac{KA}{AM} \cdot \frac{MG'}{G'A'} = \frac{1}{1} \cdot \left(-\frac{2}{1}\right) \cdot \frac{1}{2} = -1.$$

So, by Menelaus' theorem, G' lies on the median AN of triangle ABC. Similarly, G' also lies on the other medians of ABC. This proves the second claim. Combining this with the similarity from the first claim, it follows that the centroid of DEF is on line OG, namely the Euler line of ABC.

■

Problem 66. (Petrisor Neagoe, AOPS) Let $ABCD$ be a cyclic quadrilateral with circumcircle Γ and let E be an arbitrary point on the side AB. Let DE meet BC at F, DE meet again Γ at P, BP meet AF at Q, and QE meet CD at V. Prove that point V is independent of the position of E.

Solution. Let the tangent to Γ at B meet the line AD at M, and let $\mathcal{C}$ be the conic passing through A, B, C and tangent to lines MA and MB.

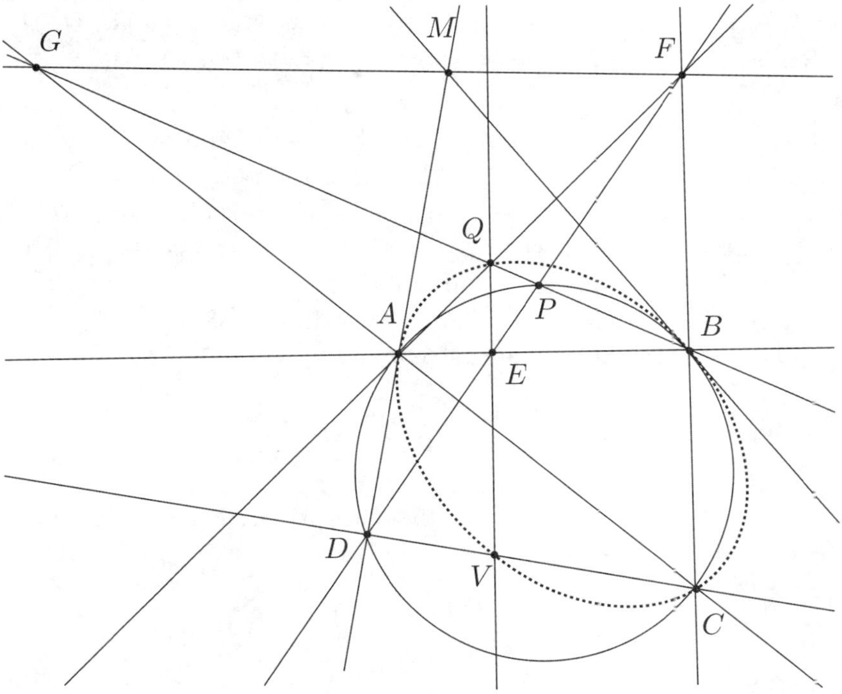

By Pascal's theorem, applied for the degenerate hexagon $BBCADP$, the intersections $M = BB \cap AD$, $F = BC \cap DP$, and $G = CA \cap PB$ are collinear. The same Pascal's theorem, this time applied for the degenerate hexagon $AAQBBC$, yields that Q lies on the conic $\mathcal{C}$. Furthermore, by Pascal's theorem in $AAQVCB$, we also get that V lies on $\mathcal{C}$; hence V is the second intersection of $\mathcal{C}$ with CD, which, in particular, shows that it is fixed. This completes the proof.

■

Problem 67. (Fedor Ivlev, RMM 2012) Let ABC be a triangle and let I and O denote its incenter and circumcenter, respectively. Let ω_A be the circle through B and C that is tangent to the incircle of the triangle ABC; the circles ω_B and ω_C are defined similarly. The circles ω_B and ω_C meet at a point A' distinct from A; the points B' and C' are defined similarly. Prove that the lines AA', BB' and CC' are concurrent at a point on the line IO.

Solution. Let γ be the incircle of the triangle ABC and let A_1, B_1, C_1 be its contact points with the sides BC, CA, AB, respectively. Let X_A be the point of contact of the circles γ and ω_A. The latter circle is the image of the former under a homothety centered at X_A. This homothety sends A_1 to a point M_A on ω_A such that the tangent to ω_A at M_A is parallel to BC. Consequently, M_A is the midpoint of the arc BC of ω_A not containing X_A. It follows that the angles M_AX_AB and M_ABC are congruent, so the triangles M_ABA_1 and M_AX_AB are similar, which gives $\frac{M_AB}{M_AX_A} = \frac{M_AA_1}{M_AB}$. Rewrite the equation as $M_AB^2 = M_AA_1 \cdot M_AX_A$ to deduce that M_A lies on the radical axis ℓ_B of B and γ. Similarly, M_A lies on the radical axis ℓ_C of C and γ.

Define the points X_B, X_C, M_B, M_C and the line ℓ_A in a similar way and notice that the lines ℓ_A, ℓ_B, ℓ_C support the sides of the triangle $M_AM_BM_C$. The lines ℓ_A and B_1C_1 are both perpendicular to AI, so they are parallel. Similarly, the lines ℓ_B and ℓ_C are parallel to C_1A_1 and A_1B_1, respectively. Consequently, the triangle $M_AM_BM_C$ is the image of the triangle $A_1B_1C_1$ under a homothety Θ. Let K be the center of Θ and let

$$k := \frac{M_AK}{A_1K} = \frac{M_BK}{B_1K} = \frac{M_CK}{C_1K}$$

be the similitude ratio. Notice that the lines M_AA_1, M_BB_1 and M_CC_1 are concurrent at K.

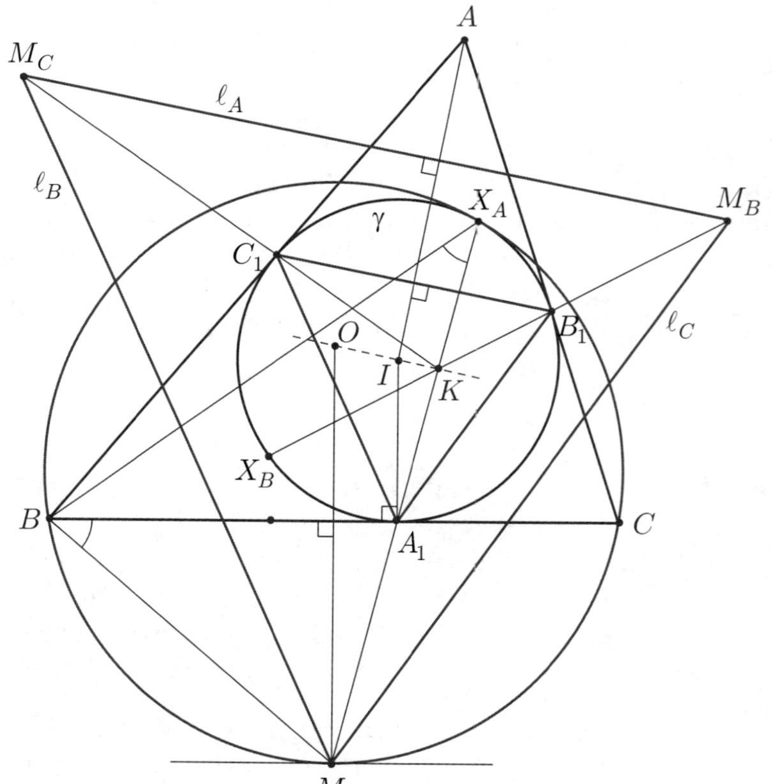

Since the points A_1, B_1, X_A, X_B are concyclic, $A_1K \cdot KX_A = B_1K \cdot KX_B$. Multiply both sides by k to get $M_AK \cdot KX_A = M_BK \cdot KX_B$ and deduce thereby that K lies on the radical axis CC' of ω_A and ω_B. Similarly, both lines AA' and BB' pass through K. Finally, consider the image O' of I under Θ. It lies on the line through M_A parallel to A_1I (and hence perpendicular to BC); since M_A is the midpoint of the arc BC, this line must be M_AO. Similarly, O' lies on the line M_BO, so $O' = O$. Consequently, the points I, K and O are collinear.

■

Comment 1. Here is an outline of an alternative approach to the first part of the solution. Let J_A be the excenter of triangle ABC opposite A. The line J_AA_1 meets γ again at Y_A; let Z_A and N_A be the midpoints of the segments A_1Y_A and J_AA_1, respectively.

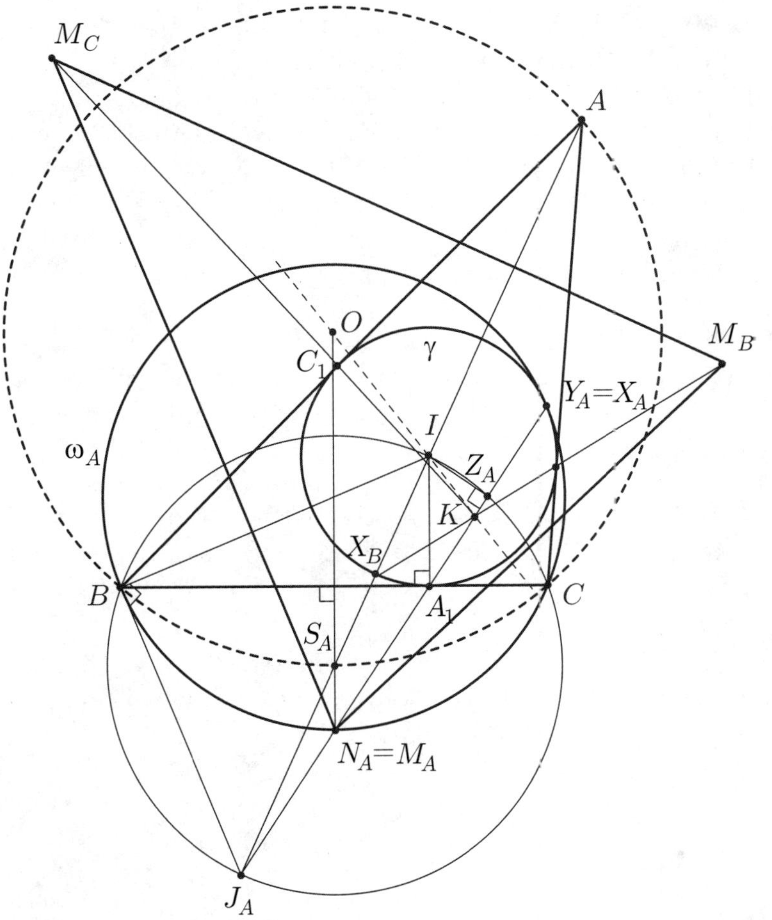

Since the segment IJ_A is a diameter in the circle BCZ_A, it follows that $BA_1 \cdot CA_1 = Z_AA_1 \cdot J_AA_1$, so $BA_1 \cdot CA_1 = N_AA_1 \cdot Y_AA_1$. Consequently, the points B, C, N_A and Y_A lie on some circle ω'_A.

It is well-known that N_A lies on the perpendicular bisector of segment BC, so the tangents to ω'_A and γ at N_A and A_1 are parallel. It follows that the tangents to these circles at Y_A coincide, so ω'_A is in fact ω_A, whence $X_A = Y_A$ and $M_A = N_A$. It is also well known that the midpoint S_A of the segment IJ_A lies both on the circumcircle ABC and on the perpendicular bisector of BC. Since S_AM_A is a midline in the triangle A_1IJ_A, it follows that $S_AM_A = \frac{r}{2}$, where r is the radius of γ (the inradius of triangle ABC). Consequently, each of the points M_A, M_B and M_C is at distance $R + \frac{r}{2}$ from O (here R is the circumradius). Now proceed as above.

Comment 2. Many steps in the first solution allows different reasonings. For instance, one may see that the lines A_1X_A and B_1X_B meet at a point K on the radical axis CC' of circles ω_A and ω_B by applying Newton's Theorem to the quadrilateral $X_AX_BA_1B_1$ (since the common tangents at X_A and X_B intersect on CC'). Then, one can conclude that $\frac{KA_1}{KB_1} = \frac{KM_A}{KM_B}$, thus obtaining that triangles $M_AM_BM_C$ and $A_1B_1C_1$ are homothetic at K (and therefore K is the radical center of ω_A, ω_B, ω_C). Finally, considering the inversion with the pole K and power equal to $KX_1 \cdot KM_A$ followed by the reflection at P, we see that the circles ω_A, ω_B, ω_C are invariant under this transform; next, the image of γ is the circumcircle of $M_AM_BM_C$ and it is tangent to all the circles ω_A, ω_B, and ω_C, hence its center is O, and thus O, I, and K are collinear.

Problem 68. (Son Hong Ta, adapter after IMO 2009) Let ABC be a triangle with circumcircle Ω. Points X and Y are on Ω such that XY meets AB and AC at D and E, respectively. Prove that the midpoints of segments XY, BE, CD, and DE are concyclic.

Solution. Let M_1, M_2, M_3, M_4 be the midpoints of segments XY, DE, BE, CD. Let A_1, B_1, C_1 be the second intersections of the lines AM_1, BM_1, CM_1 with Ω. Furthermore, let $C_2 = C_1A_1 \cap XY$, $B_2 = B_1A_1 \cap XY$, and $A_2 = CB_2 \cap BC_2$.

Since B_2, C_2 and M are collinear and C_1, C, A_2, B, B_1 all lie on the same circle, we can apply the converse of Pascal's theorem on hexagon $A_1C_1CA_2BB_1$ to get that A_2, lies on Ω. Thus

$$\begin{aligned} \angle BAC &= \angle BA_2C \\ &= 180^\circ - \angle C_2BC - \angle B_2CB \\ &= \angle BC_2E + \angle CB_2D - 180^\circ. \end{aligned}$$

On the other hand, the Butterfly Theorem yields $DM_1 = M_1B_2$ and $EM_1 = M_1C_2$. Hence from SAS similarity we see that $M_1M_3 \| C_2B$,

$M_1M_2 \| B_2C$. It follows that

$$\begin{aligned}\angle BAC &= \angle BC_2E + \angle CB_2D - 180^\circ \\ &= \angle M_3M_1E + \angle M_4M_1D - 180^\circ \\ &= \angle M_4M_1M_3.\end{aligned}$$

Also since $M_2M_3 \| AB, M_2M_4 \| AC$, we have $\angle BAC = \angle M_4M_2M_3$; therefore $\angle M_4M_2M_3 = \angle M_4M_1M_3$, from which it follows thatM_1, M_2, M_3, M_4 are concyclic, as claimed. This completes the proof.

■

Problem 69. (Droz-Farny's Line Theorem) If two perpendicular straight lines are drawn through the orthocenter of a triangle, they intercept a segment on each of the sidelines. Prove that the midpoints of these three segments are collinear.

Solution. We will denote by A_1, B_1, C_1, and A_2, B_2, C_2 the intersections points of the two perpendicular lines ℓ_1, ℓ_2 with the sidelines BC, CA, and AB, respectively. The Droz-Farny line theorem states that the midpoints A_3, B_3, C_3 of the segments A_1A_2, B_1B_2, C_1C_2 are collinear.

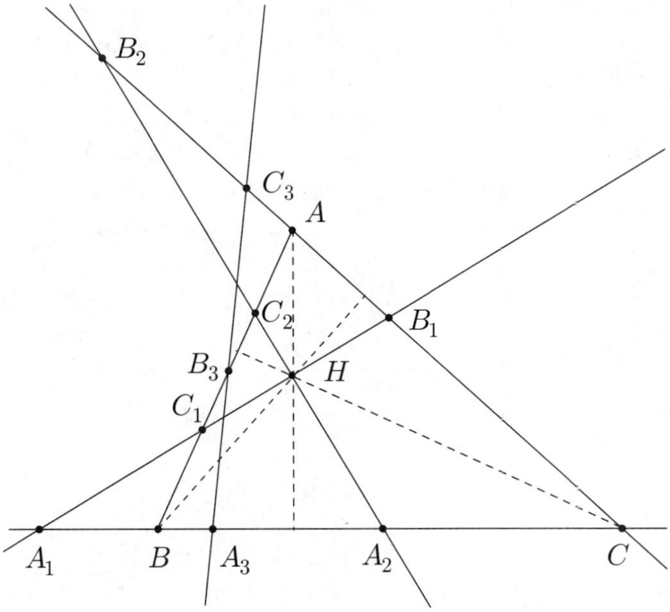

Let H_a, H_b, H_c be the reflections of H across BC, CA, AB, respectively. The lines A_1H_a, B_1H_b, C_1H_c are then the reflections of ℓ_1 with respect to BC, CA, AB, respectively; so, by Collings' theorem (see the proof of Problem 48), they meet at some point S, lying on the circumcircle of ABC.

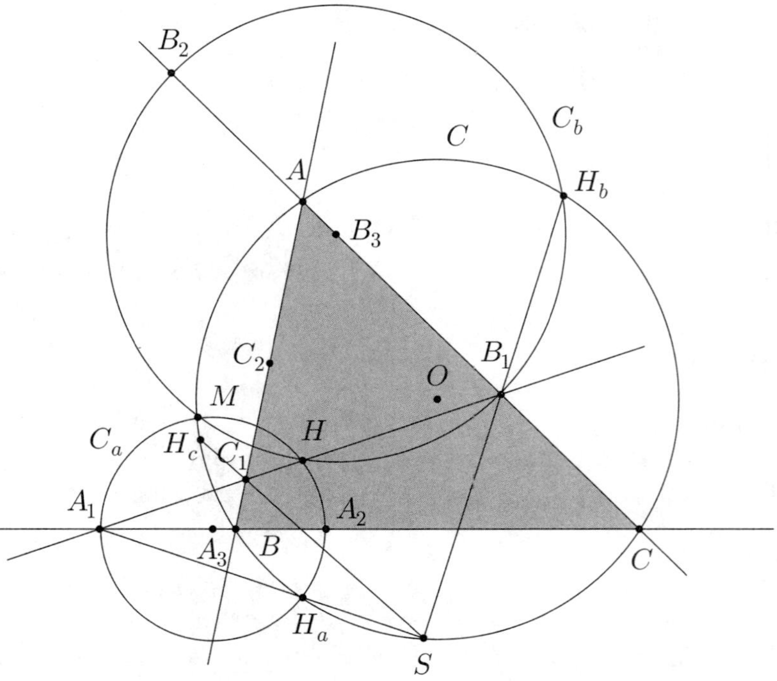

Now, assume without loss of generality that A_1 is on the extension of BC and that B_1, C_1 are inside CA and AB, respectively. By Miquel's Theorem, applied in triangle SA_1B_1, for the points H, H_a, H_b, lying on A_1B_1 ,A_1S, B_1S, respectively, we get that the circumcircles of triangles A_1HH_a, B_1HH_b and SH_aH_b are concurrent. But the circumcircles of triangles A_1HH_a and B_1HH_B are just the circles of diameters A_1A_2 and B_1B_2, whereas the circumcircle of SH_aH_b is just the circumcircle of ABC. So, by doing the same thing for NXZ (Miquel's Theorem), we get that the circles of diameters A_1A_2, B_1B_2, C_1C_2 and the circumcircle of ABC are all concurrent. But this just means that the circles of diameters A_1A_2, B_1B_2, C_1C_2 have two common points (they already had the common point H). Thus, in particular, their centers A_3, B_3, C_3 are collinear, which is exactly what we wanted to prove. This completes the proof of the Droz-Farny Line Theorem.

■

Comment 1. The Droz-Farny line theorem was discovered by Arnold Droz-Farny in 1899 and, despite the simple configuration, the result has quite a history of very complicated computational proofs. The synthetic solution we presented above appeared only relatively recently, in 2004, and is due to Jean-Louis Ayme. Nonetheless, a lot of generalizations and related results have appeared since then. For instance, one month after Ayme's article, Floor van Lamoen conjectured the following generalization:

Generalization 1. If the midpoints of the intercepted segments are replaced by three points A_3, B_3, C_3 dividing into the same ratio the corresponding segments A_1A_2, B_1B_2, and C_1C_2, then A_3, B_3, C_3 remain collinear.

We prove this result below using vectors.

Proof. Denote by e, f the lines through the orthocenter H that are parallel to AB, AC, respectively. Furthermore, denote by x, y the lines through the vertex A that are parallel to the lines ℓ_1, ℓ_2, and let X, Y be the intersection points of the sideline BC with x, and y, respectively.

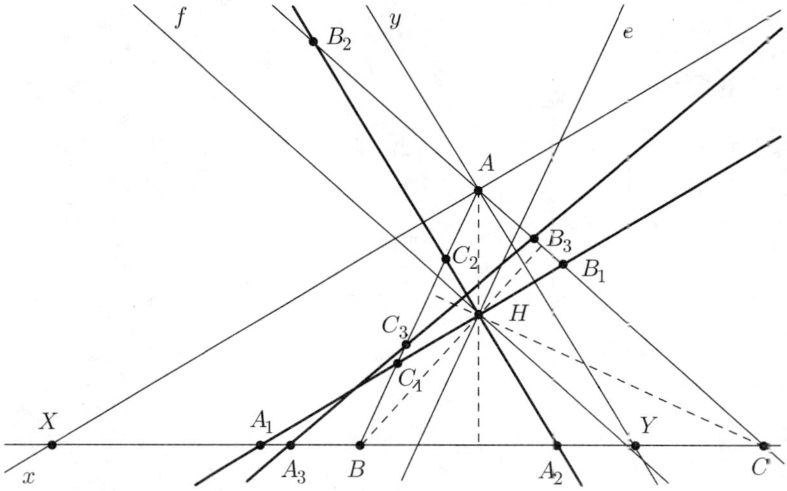

Since the pencil $(HC_1,\ HC_2,\ HB,\ e)$ is the image of $(HB_2,\ HB_1,\ f,\ HC)$ under the rotation $\Psi(H, +90°)$,

$$\frac{BC_1}{BC_2} = \frac{CB_1}{CB_2} \quad \text{if and only if} \quad \frac{BC_1}{CB_1} = \frac{BC_2}{CB_2},$$

and thus, by multiplying with $\frac{AC}{AB}$,

$$\frac{C_1B}{AB} \cdot \frac{AC}{B_1C} = \frac{C_2B}{AB} \cdot \frac{AC}{B_2C}.$$

On other hand, since

$$\frac{C_1B}{AB} = \frac{A_1B}{XB},\quad \frac{AC}{B_1C} = \frac{XC}{A_1C},\quad \frac{C_2B}{AB} = \frac{A_2B}{YB},\quad \frac{AC}{B_2C} = \frac{YC}{A_2C},$$

it follows that

$$\frac{A_1B}{A_1C} : \frac{XB}{XC} = \frac{A_2B}{A_2C} : \frac{YB}{YC},$$

which is equivalent with the congruence of the pencils $(B,\ C,\ A_1,\ X)$ and $(B,\ C,\ A_2,\ Y)$. By intersecting now $(AB,\ AC,\ AA_1,\ AX)$ with ℓ_1 and $(AB,\ AC,\ AA_2,\ AY)$ with ℓ_2, we deduce that

$$\frac{C_1A_1}{C_1B_1} = \frac{C_2A_2}{C_2B_2},$$

the two degenerated triangles $A_1B_1C_1$ and $A_2B_2C_2$ being similar.

For a point P denote by $\mathbf{P}$ the vector $\overrightarrow{XP}$, where X is a fixed point in plane of triangle ABC. Since $\frac{C_1A_1}{C_1B_1} = \frac{C_2A_2}{C_2B_2}$, there exist two real numbers k, l, satisfying $k + l = 1$, such that

$$\mathbf{C_1} = k\mathbf{A_1} + l\mathbf{B_1}, \quad \mathbf{C_2} = k\mathbf{A_2} + l\mathbf{B_2}.$$

On other hand, since A_3, B_3, C_3 divide the segments A_1A_2, B_1B_2, and C_1C_2, respectively, into the same ratio, there exist two real numbers u, v, satisfying $u + v = 1$, such that

$$\mathbf{A_3} = u\mathbf{A_1} + v\mathbf{A_2}, \quad \mathbf{B_3} = u\mathbf{B_1} + v\mathbf{B_2}, \quad \mathbf{C_3} = u\mathbf{C_1} + v\mathbf{C_2}.$$

Therefore,

$$\begin{aligned} \mathbf{C_3} = u\mathbf{C_1} + v\mathbf{C_2} &= u\left(k\mathbf{A_1} + l\mathbf{B_1}\right) + v\left(k\mathbf{A_2} + l\mathbf{B_2}\right) \\ &= k\left(u\mathbf{A_1} + v\mathbf{A_2}\right) + l\left(u\mathbf{B_1} + v\mathbf{B_2}\right) \\ &= k\mathbf{A_3} + l\mathbf{B_3}. \end{aligned}$$

According to the fact that $k + l = 1$, this implies that the points A_3, B_3, C_3 are collinear. This completes the proof.

Comment 2. Let us now give another generalization of a different appeal.

Generalization 2. Let P be a point in plane and let γ be a line passing through it. If A', B', C' are the points where the reflections of lines PA, PB, PC with respect to γ meet the sides BC, CA, AB, respectively, then A', B', C' are collinear.

If this looks familiar, it is because it is in fact a problem from USAMO 2012. We refer to 107 Geometry Problems from the AwesomeMath Year-Round Program for a proof using Menelaus' Theorem. Just for fun, we include a second proof using isogonal conjugates.

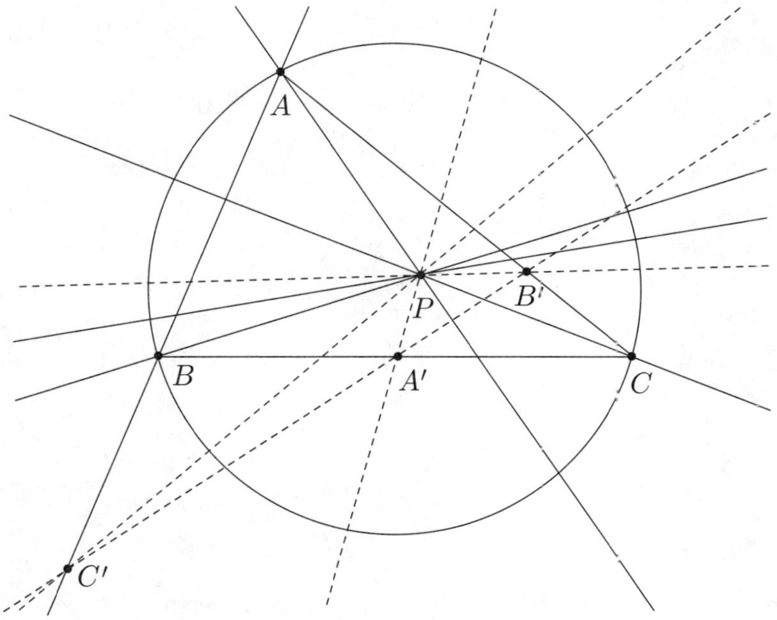

Proof of Generalization 2. We split the solutions into two parts according to the position of P with respect to the circumcircle of ABC.

First, if P lies on the circumcircle then we can easily see that P is actually the Miquel point of the triplet of points A', B', C' with respect triangle ABC. Indeed, since $\angle A'BC' = \angle CBA = \angle CPA = \angle A'PC'$, it follows that the points P, A', B, C' are concyclic, and similarly for P, A, B', C', and P,A', B', C. Hence, $\angle CA'B' = \angle CPB' = \angle BPC' = \angle BA'C'$, so A', B', C' are collinear.

Now, if P is not on the circumcircle of ABC, then let Q be its isogonal conjugate with respect to triangle ABC (which is not degenerate - i.e. at infinity - as P does not lie on the circumcircle).

Claim. Furthermore, let Q' be the isogonal conjugate of P with respect to triangle $AB'C'$. Then, $Q = Q'$.

Proof of Claim. To begin, note that

$$\begin{aligned}\angle BQC &= \angle BAC + \angle CPB \text{ (since } P \text{ and } Q \text{ are isogonal conjugates in } \triangle ABC)\\ &= \angle C'AB' + \angle B'PC'\\ &= \angle C'Q'B' \text{ (since } P \text{ and } Q \text{ are isogonal conjugates in } \triangle AB'C').\end{aligned}$$

Denote by X, Y, Z the reflections of P in the sides BC, CA, AB, and by X' its reflection in the side $B'C'$ of triangle $AB'C'$. Then, $\angle ZXY = \angle BQC$ (because QC is orthogonal to XY and QB is orthogonal to XZ), whereas $\angle ZX'Y = \angle C'Q'B'$ (because $Q'B'$ is orthogonal to $X'Y$ and $Q'C'$ is orthogonal $X'Z$), so, since $\angle C'Q'B' = \angle BQC$, we get $\angle ZXY = \angle ZX'Y$. From this it follows that X, Y, Z, and X' are concyclic. However, the center of the XYZ-circle is Q while the center of the $X'YZ$-circle is Q' (we make use of the well known fact that the circumcenter of the triangle determined by the reflections of a point across the sidelines of a triangle is precisely the isogonal conjugate of the point with respect to that triangle; for a proof, see for example the classic R. A. Jonhson, *Advanced Euclidean Geometryl*). Thus, Q and Q' do indeed coincide. This proves our Claim.

Now, in a similar way, we can deduce that Q is also the isogonal point of P with respect to triangles $A'BC'$ and $A'B'C$. Therefore,

$$\begin{aligned}\angle BC'A' &= \angle AC'A' = \angle AC'P + \angle PC'Q + \angle QC'A' \\ &= \angle QC'B' + \angle PC'Q + \angle BC'P \\ &= \angle BC'B' = \angle AC'B'.\end{aligned}$$

This means that A', B', C' are collinear, thus completing the proof.

Why is Generalization 2 a generalization of the Droz-Farny line theorem? This is not completely trivial to notice. We reformulate the USAMO problem as follows:

Reformulation. Let γ and δ be lines passing through a given point P in the plane of a triangle ABC. Let A', B', C' and A'', B'', C'' be the intersections of γ and δ with BC, CA, AB, respectively. Furthermore, let X be the intersection of BC with the reflection of AP into the internal angle bisector of $angle A'PA''$, and similarly, define Y and Z. Then points X, Y, Z are collinear.

Of course, this new statement involves the redundant definition of A' B', C', A'', B'', C'' , but it points out the remarkable fact that the Droz-Farny line theorem is a special case of the USAMO configuration, where we take P to be the orthocenter of ABC and the lines γ and δ to be perpendicular. The lines AP, BP, CP, besides being the altitudes in ABC, become the altitudes in triangles $A'PA''$, $B'PB''$, $C'PC''$, respectively; hence, their reflections in the corresponding internal angle bisectors pass through the circumcenters of triangles $A'PA''$, $B'PB''$, $C'PC''$ (as the orthocenter and the circumcenter are isogonal conjugates), which are precisely the midpoints of the interceptions from Droz-Farny.

Problem 70. (Christopher Bradley, Crux Mathematicorum) Suppose that ABC is an equilateral triangle and that P is a point in the plane of ABC.

The perpendicular from P to BC meets AB at X, the perpendicular from P to CA meets BC at Y, and the perpendicular from P to AB meets CA at Z.

(a) If P is in the interior of triangle ABC, prove that the area of XYZ is not greater than the area of ABC.

(b) If P lies on the circumcircle of ABC, prove that X, Y, and Z are collinear.

Solution. (a) Let D, E, F be the points of intersection of PX with BC, PY with CA, PZ with AB, respectively. Since $XD \perp BC$ and $\angle ABC = 60°$, we have $\angle PXF = \angle DXB = 30°$. Thus, $PF = PX \sin 30° = \frac{1}{2}PX$; that is, $PX = 2PF$. Similarly, $PY = 2PD$ and $PZ = 2PE$.

Since $\angle DPZ = \angle XPF = 60°$, $\angle EPX = \angle YPD = 60°$, and $\angle FPY = \angle ZPE = 60°$, we have $\angle XPY = \angle FPD = 120°$. Hence,

$$\frac{[XPY]}{[FPD]} = \frac{PX \cdot PY}{PF \cdot PD} = \frac{2PF \cdot 2PD}{PF \cdot PD} = 4.$$

Therefore, $[XPY] = 4[FPD]$. Similarly, we have $[YPZ] = 4[DPE]$ and $[ZPX] = 4[EPF]$. Thus,

$$\begin{aligned}[XYZ] &= [XPY] + [YPZ] + [ZPX] \\ &= 4\left([FPD] + [DPE] + [EPF]\right) \\ &= 4[DEF].\end{aligned}$$

Now, let $PD = x$, $PE = y$, and $PF = z$, and let k be the altitude of ABC. As is well known, we have $x + y + z = k$. Since

$$\angle FPD = \angle DPE = \angle EPF = 120°,$$

we have

$$\begin{aligned}[DEF] &= [DPE] + [EPF] + [FPD] \\ &= \frac{1}{2}xy \sin 120° + \frac{1}{2}yz \sin 120° + \frac{1}{2}zx \sin 120° \\ &= \frac{\sqrt{3}}{4}(xy + yz + zx).\end{aligned}$$

Since

$$(x + y + z)^2 - 3(xy + yz + zx) = x^2 + y^2 + z^2 - (xy + yz + zx) \geq 0,$$

we see that $xy + yz + zx \leq \frac{1}{3}k^2$. Thus,

$$[DEF] \leq \frac{\sqrt{3}}{4} \cdot \frac{1}{3}k^2 = \frac{\sqrt{3}}{12}k \cdot \frac{\sqrt{3}}{2}BC = \frac{1}{4}[ABC].$$

Therefore, $[XYZ] \leq [ABC]$, as desired.

(b) Without loss of generality, we may assume that P is a point on the minor arc BC. Since $XD \perp BC$, we have $\angle PXB = \angle DXB = 30^\circ$. Similarly, $\angle PYB = \angle EYC = 30^\circ$ and $\angle PZC = \angle FZA = 30^\circ$. Since $\angle PXB = \angle PYB$, we see that B, P, Y, X are concyclic. Thus $\angle PYX = \angle PBF$. Since $\angle PYB = \angle PZC$, we see that P, Z, C, Y are concyclic. Thus, $\angle PYZ = \angle PCZ$.

Since A, B, P, C are concyclic, we have $\angle PBF = \angle PCA$. Hence, it follows that $\angle PYX = \angle PCA$. Using $\angle PYZ = \angle PCZ$, we now write

$$\angle PYX + \angle PYZ = \angle PCA + \angle PCZ = 180^\circ.$$

Therefore, X, Y, Z are collinear.

■

Problem 71. (Vasily Mokin and Fedor Ivlev, RMM 2010) A triangle ABC is inscribed in a circle ω. A variable line ℓ chosen parallel to BC meets segments AB, AC at points D, E, respectively, and meets ω at points K, L (where D lies between K and E). Circle γ_1 is tangent to the segments KD and BD and also tangent to ω, while circle γ_2 is tangent to the segments LE and CE and also tangent to ω. Determine the locus, as ℓ varies, of the meeting point of the common inner tangents to γ_1 and γ_2.

Solution. Let P be the meeting point of the common inner tangents to γ_1 and γ_2. Also, let b be the angle bisector of $\angle BAC$. Since $KL \| BC$, b is also the angle bisector of KAL. Let $\mathcal{T}$ be the composition of the symmetry $\mathcal{S}$ with respect to b and the inversion $\mathcal{I}$ of center A and power $AK \cdot AL$ (it is readily seen that $\mathcal{S}$ and $\mathcal{I}$ commute, so since $\mathcal{S}^2 = \mathcal{I}^2 = \text{id}$, then also $\mathcal{T}^2 = \text{id}$, the identity transformation).

Transformation $\mathcal{T}$ interchanges the following elements: point $K \longleftrightarrow$ point L, line $KL \longleftrightarrow$ circle ω, ray $AB \longleftrightarrow$ Ray AC, point $B \longleftrightarrow$ point E, point $C \longleftrightarrow$ point D, segment $BD \longleftrightarrow$ segment EC, arc $BK \longleftrightarrow$ segment EL, arc $CL \longleftrightarrow$ segment DK.

Let O_1 and O_2 be the centers of circles γ_1 and γ_2. Since the circles γ_1 and γ_2 are determined by their construction (in a unique way), they are interchanged by $\mathcal{T}$, therefore the rays AO_1 and AO_2 are symmetric with respect to b. Denote by ρ_1 and ρ_2 the radii of γ_1 and γ_2. Since $\angle O_1AB = \angle O_2AC$, we have $\frac{\rho_1}{\rho_2} = \frac{AO_1}{AO_2}$. On the other hand, from the definition of P we have

$$\frac{O_1P}{O_2P} = \frac{\rho_1}{\rho_2} = \frac{AO_1}{AO_2};$$

this means that AP is the angle bisector of $\angle O_1AO_2$ and therefore of $\angle BAC$.

The limiting, degenerate, cases are when the parallel line passes through A or when P coincides with A; respectively when the parallel line is BC and when P coincides with the foot $A' \in BC$ of the angle bisector of $\angle BAC$ (or any other point on BC). By continuity, any point P on the open segment AA' is obtained for some position of the parallel, and therefore the locus is the open segment AA' of the angle bisector b of $\angle BAC$.

■

Problem 72. (Nguyen Hoang Son, AOPS Forum) Let $ABCD$ be a cyclic quadrilateral with circumcircle Γ. Let E be the intersection of AB and CD, F the intersection of AD and BC, and M, N the midpoints of AC and BD, respectively. Prove that EF is tangent to the circumcircle of triangle MNF.

Solution. The proofs for this magnificent property of cyclic quadrilaterals requires two important Lemmas: i) the famous Newton-Gauss theorem that the midpoints of the diagonals of a complete quadrilateral are collinear and ii) the following result about harmonic divisions.

Lemma. The quadruplet of collinear points (A, C, B, D) is harmonic if and only if

$$MA \cdot MB = MC^2,$$

where M is the midpoint of the segment CD.

Proof. Let P be a point on the semicircle with diameter CD, not lying on the line CD. We have that $MP = MC = MD$ and that $PC \perp PD$. In particular, we have that

$$MP^2 = MC^2 = MA \cdot MB,$$

which is equivalent with the line MP being tangent to the circumcircle of triangle PAB. Thus, we get that

$$MA \cdot MB = MC^2$$

is equivalent with $\angle MPB = \angle PAB$. However, recall that (A, C, B, D) is harmonic if and only if $\angle APC = \angle BPC$. However, triangle MPC is isosceles, so this happens if and only if $\angle MPB = \angle PAB$. This proves the Lemma.

□

For sake of completeness, before going to the proof of the problem, let us make a sidenote with the Newton-Gauss theorem and give a neat proof.

Newton-Gauss Theorem. Let $ABCD$ be a (convex) quadrilateral and let E be the intersection of AB and CD and F be the intersection of AD and BC. Then, the midpoints of segments AC, BD and EF are collinear.

The line determined by these midpoints is known is in fact usually called the *Gauss line* of quadrilateral $ABCD$, but here we shall continue calling it the Newton-Gauss line. Standards proofs use multiple applications of Menelaus' theorems or area calculations. Normally, we would have left such a well known theorem unproven, but, after Solution 8, we would like include a quick proof with Pappus' theorem that is quite beautiful and not (yet) that popular.

Proof. Let G, H, I be the midpoints of segments BD, AC, EF, and let J, K, L be the midpoints of segments DC, CE, DE, respectively. Furthermore, let M be the intersection of HJ with GK, N the intersection of HL with IK, and O the intersection of IJ with GL.

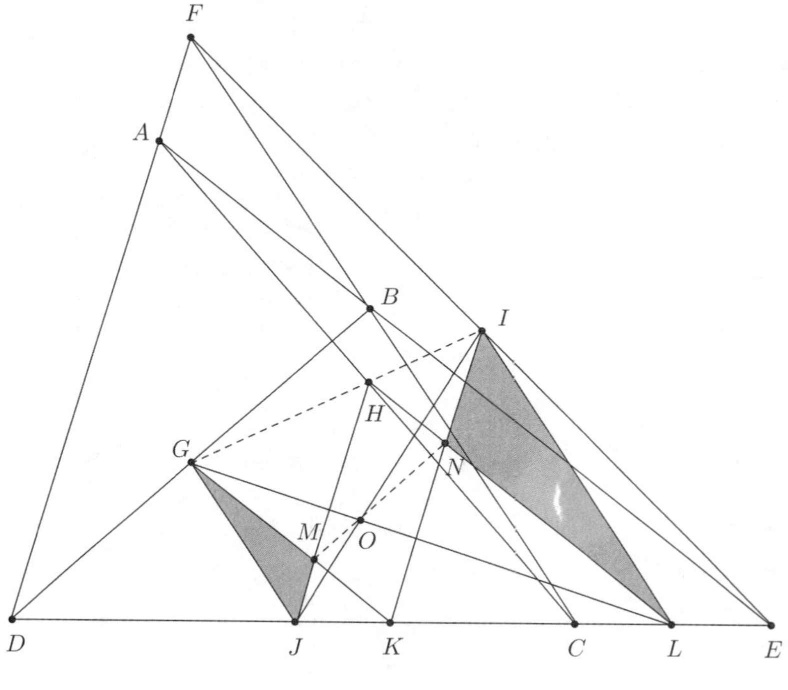

Note that GK is the $D-$midline in triangle DBE, so $GK \| BE$, or $GM \| BE$. Also, HL is the $C-$midline in triangle CAE, so $HL \| AE$, or $NL \| BE$. In particular, it follows that $GM \| NL$. Analogously, we get that $GJ \| IL$ and $MJ \| IN$, which implies that triangles GMJ And LNI are homothetic. In particular, this means that the lines GL, MN, and JI are concurrent, or in other words that point O lies on the line MN. This creates for us the exact setting to apply Pappus' theorem: the points M, O, N are collinear and the points L, K, J are collinear; hence, the intersections $G = MK \cap OL$, $I = OJ \cap NK$, $H = MJ \cap NL$ are collinear, which is precisely what we wanted to show. This completes the proof.

□

Finally returning to the problem, let τ_A, τ_B, τ_C, τ_D of Γ at the vertices A, B, C, D, respectively. Let G be the intersection of τ_A and τ_C, and let H be the intersection of τ_B and τ_D. Furthermore, let P be the intersection of AC and BD.

Recall the well known fact that EF is the polar of P with respect to Γ. [If this sounds new, just take the intersections X, Y of EP with BC and AD, respectively, and note that the quadruplets (F, B, X, C) and (F, A, Y, D) are harmonic, since lines EX, BD, CA are concurrent cevians in triangle EBC; this proves that X and Y lie on the polar of F with respect to Γ, and so EP is the polar of F with respect to Γ. Analogously, FP is the polar of E with respect to Γ; hence, P is the pole of EF with respect to Γ, as claimed.

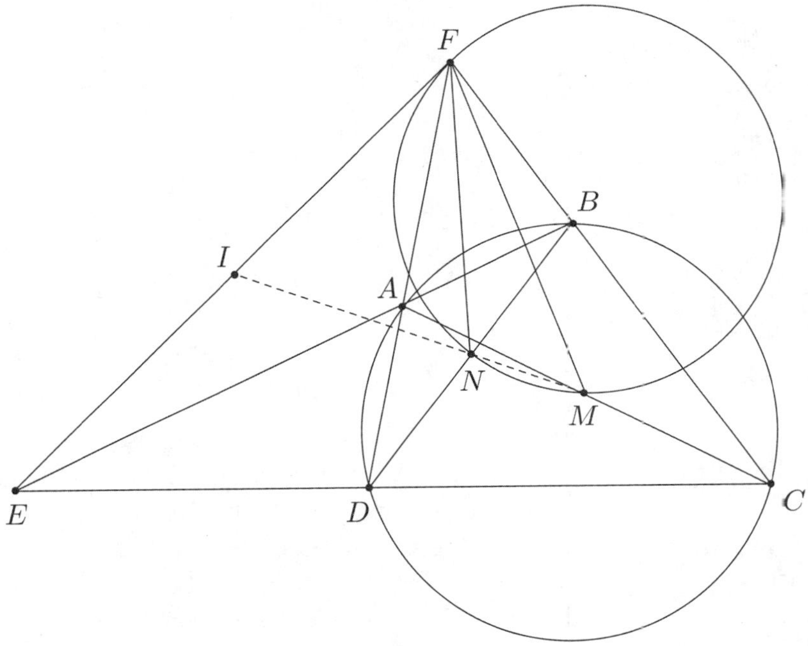

On the other hand, AC is the polar of $G = \tau_A \cap \tau_C$ and BD is the polar of $H = \tau_B \cap \tau_D$; thus G and H both lie on the polar of $P = AC \cap BD$. Therefore, the points E, F, G, H are all colinear. Moreover, these points form a harmonic division. Indeed, since the tangents at B and D with respect to Γ meet at H, note that CH is the $C-$symmedian of triangle CBD, whereas CG is the C-tangent with respect to the circumcircle Γ. Hence (CD, CH, CB, CG) is a harmonic pencil, which intersected with line EF, tells us that (E, H, F, G) is a harmonic pencil as claimed.

Let I be the midpoint of segment EF. By the Newton-Gauss theorem, the points M, N and I are collinear. Quadruplet (E, H, F, G) is harmonic, so the Lemma tells us that

$$IF^2 = IG \cdot IH.$$

On the other hand, if O and R denote the circumcenter and circumradius of quadrilateral $ABCD$,

$$OM \cdot OG = ON \cdot OH = R^2,$$

hence quadrilateral $GHMN$ is cyclic. By Power of Point, this also means however that

$$IN \cdot IM = IG \cdot IH = IF^2,$$

which enables us to conclude that IF is tangent to the circumcircle of triangle MNF. It follows that EF is tangent to the circumcircle of MNF, as claimed. ■

Problem 73. (Toshio Seimiya, Crux Mathematicorum) Consider triangle ABC, and three squares $BCDE$, $CAFG$ and $ABHI$ constructed on its sides, outside the triangle. Let XYZ be the triangle enclosed by the lines EF, DI and GH. Prove that $[XYZ] \leq (4 - 2\sqrt{3})[ABC]$.

Solution. Surprisingly, the $2\sqrt{3}[ABC]$ term comes from the celebrated

Weitzenbock Inequality. In a triangle ABC with side lengths $a = BC$, $b = CA$, $c = AB$, we have that

$$a^2 + b^2 + c^2 \geq 4\sqrt{3}[ABC].$$

This beautiful inequality is due to Weitzenbock, but has also appeared in numerous contests, the most famous being the IMO of 1961. In [29], one can find eleven proofs. We shall omit doing it here, since, as a matter of fact, we proved something far stronger in Problem 109, which generalized not only Weitzenbock's inequality but also the Hardwiger-Finsler improvement

$$a^2 + b^2 + c^2 \geq 4\sqrt{3}[ABC] + (a-b)^2 + (b-c)^2 + (c-a)^2.$$

Returning to the problem, let A', B', C' be the centers of the squares $BCDE$, $CAFG$, $ABHI$, respectively. We shall use directed areas so that, for example, $[A'BC'] > 0$ when $\angle B > 90°$. We split the proof into three big claims.

Claim 1. $[AB'C'] + [A'BC'] + [A'B'C] = \frac{a^2+b^2+c^2}{8}$.

Claim 2. $[A'B'C'] = [ABC] + \frac{a^2+b^2+c^2}{8}$.

Claim 3. $[XYZ][A'B'C'] = [ABC]^2$.

Indeed, if we manage to prove these three results, we can then employ Weitzenbock's inequality to get that

$$[A'B'C'] = [ABC] + \frac{a^2+b^2+c^2}{8} \geq \left(1 + \frac{\sqrt{3}}{2}\right)[ABC].$$

This combined with the identity from Claim 3 then gives

$$[XYZ]\left(1 + \frac{\sqrt{3}}{2}\right)[ABC] \leq [XYZ][A'B'C'] = [ABC]^2.$$

Hence,

$$[XYZ] \leq \left(\frac{1}{1+\frac{\sqrt{3}}{2}}\right)[ABC] = (4-2\sqrt{3})[ABC],$$

as claimed. It thus remains to show that Claims 1, 2, and 3 are true.

Proof of Claim 1. A spiral similarity with center A and ratio of magnification $\sqrt{2}$ takes triangle $AB'C'$ to triangle AFB; hence $[AB'C'] = \frac{1}{2}[AFB]$. Similarly, $[CA'B'] = \frac{1}{2}[CBG]$. Consequently,

$$[AB'C'] + [A'B'C] = \frac{1}{2}[AFB] + \frac{1}{2}[CBG] = \frac{1}{4}[CAFG] = \frac{b^2}{4}.$$

Similarly,

$$[AB'C'] + [A'BC'] = \frac{c^2}{4}, \quad \text{and} \quad [A'BC'] + [A'B'C] = \frac{a^2}{4}.$$

Add the three equalities for the result.

[Alternatively, one can use the Law of Cosines to prove the claim. Indeed, write

$$\begin{aligned} [AB'C'] &= \frac{1}{2}AB' \cdot AC' \sin\angle B'AC' \\ &= \frac{1}{2}\frac{AC}{\sqrt{2}} \cdot \frac{AB}{\sqrt{2}}\sin(90^\circ + A) \\ &= \frac{1}{4}bc\cos A \\ &= \frac{b^2+c^2-a^2}{8} \end{aligned}$$

with similar expressions for $[A'BC']$ and $[A'B'C]$.]

Proof of Claim 2. Write

$$\begin{aligned}[A'B'C'] &= [ABC]+[A'BC]+[AB'C]+[ABC']-([AB'C']+[A'BC']+[A'B'C]) \\ &= [ABC]+\frac{a^2}{4}+\frac{b^2}{4}+\frac{c^2}{4}-\frac{a^2+b^2+c^2}{8} \\ &= [ABC]+\frac{a^2+b^2+c^2}{8},\end{aligned}$$

which is precisely the identity we need.

Proof of Claim 3. Since C' and B' are midpoints of two sides of AHG, we have that GH is parallel to $B'C'$. The analogous statements for $A'B'$ and $A'C'$ imply that triangles XYZ and $A'B'C'$. Let $XY = \lambda A'B'$; then

$$[XYZ] = \lambda^2[A'B'C'].$$

Hence, the area of the region between triangles $A'B'C'$ and XYZ is

$$\begin{aligned}(1-\lambda^2)[A'B'C'] &= [A'B'C'] - [XYZ] \\ &= ([A'B'X]+[XB'Y]) + ([B'C'Y]+[YC'Z]) \\ &\quad + ([C'A'Z]+[ZA'X]) \\ &= (1+\lambda)([A'B'X]+[B'C'Y]+[C'A'Z]) \\ &= (1+\lambda)([A'CB']+[B'AC']+[C'BA']).\end{aligned}$$

Therefore,

$$(1-\lambda)[A'B'C'] = [A'CB']+[B'AC']+[C'BA'],$$

and

$$\lambda = \frac{[A'B'C'] - ([A'CB']+[B'AC']+[C'BA'])}{[A'B'C']} = \frac{[ABC]}{[A'B'C']}.$$

Finally, plug this expression for λ back to the first equation to get the desired identity. This completes the proof for Claim 3 and the problem. ∎

Comment. We end the discussion with a generalization, which we leave for the reader.

Generalization. Consider a triangle ABC and three rectangles $BCDE$, $CAFG$, and $ABHI$ erected outside the triangle such that the side perpendicular to the corresponding side of the triangle is μ times as long. Let XYZ be the triangle enclosed by the lines EF, DI, and GH; then

$$[XYZ] \le (\sqrt{3}\mu - 1)^2[ABC].$$

Moreover, when $\mu \neq \frac{1}{\sqrt{3}}$ equality holds if and only if triangle ABC is equilateral; when $\mu = \frac{1}{\sqrt{3}}$, the three lines EF, DI, and GH are concurrent.

Problem 74. (Floor van Lamoen, American Mathematical Monthly) A triangle is divided by its three medians into six smaller triangles. Show that the circumcenters of these smaller triangles lie on a circle.

Solution 1. We highlight three beautiful Lemmas about harmonic quadrilaterals that will come together to yield a magnificent proof.

Lemma 1. Let ABC be a triangle and let M be a point on the arc BC of the circumcircle of triangle ABC, which does not contain vertex A. Then, AM is the A-symmedian of triangle ABC if and only if quadrilateral $ABMC$ is harmonic.

Proof. Let AM intersect the side BC at X. The line AM is the A-symmedian of triangle ABC if and only if

$$\frac{XB}{XC} = \frac{AB^2}{AC^2}.$$

However, by the Ratio Lemma, applied in triangle MBC,

$$\begin{aligned} \frac{XB}{XC} &= \frac{MB}{MC} \cdot \frac{\sin \angle XMB}{\sin \angle XMC} \\ &= \frac{MB}{MC} \cdot \frac{\sin \angle ACB}{\sin \angle ABC} \\ &= \frac{MB}{MC} \cdot \frac{AB}{AC}, \end{aligned}$$

where the latter equality holds from the Law of Sines in triangle ABC. It follows that

$$\frac{XB}{XC} = \frac{AB^2}{AC^2} \text{ if and only if } \frac{MB}{MC} = \frac{AB}{AC},$$

which proves that AM is the A-symmedian of triangle ABC if and only if quadrilateral $ABMC$ is harmonic, as claimed.

□

Lemma 2. Let ABC be a triangle and let M be an arbitrary point in its plane. Let A_1, B_1, C_1 be the orthogonal projections of M onto the sides BC, CA, AB, respectively. Then, A_1 is the midpoint of segment B_1C_1 if and only if quadrilateral $ABMC$ is harmonic.

Proof. If A_1 is the midpoint of segment B_1C_1, then in particular it lies on B_1C_1, so by Simson's theorem, point M lies on the circumcenter of triangle

ABC. Moreover, triangles MBC and MC_1B_1 are similar; therefore, if K is the midpoint of side BC,

$$\angle KMB = \angle A_1MC_1 = \angle A_1BC_1 = \angle CMA.$$

It follows that AM is the A-symmedian of triangle MBC; thus, Lemma 1, applied for triangle MBC, yields that quadrilateral $ABMC$ is harmonic. For the converse, we just reverse the above reasoning.

□

Lemma 3. Let $ABMC$ be a harmonic quadrilateral. Through M, draw two parallel lines to AB, AC, so that they intersect AC and AB at points P and Q, respectively. Then, quadrilateral $BCPQ$ is cyclic.

Proof. Denote by J the intersection of the tangents at the vertices B and C with respect to the circumcircle of quadrilateral $ABMC$. Quadrilateral $APMQ$ is a parallelogram, and therefore PQ passes through the midpoint N of segment AM. Points O, B, C, J and N thus all lie on a circle with diameter OJ, so

$$\angle BNJ = \angle BCJ = \angle A = \angle BQM.$$

Therefore, quadrilateral $BMNQ$ is cyclic, which implies

$$\angle AQN = \angle BMN = \angle BCA.$$

It follows that quadrilateral $BCPQ$ is cyclic, which completes our proof.

□

Finally returning to our problem, let us introduce some notation. Let the given triangle ABC, let AM, BN, CP be its medians, and let O_1, O_2, O_3, O_4, O_5, O_6 be the circumcenters of triangles BGM, CGM, CGN, AGN, AGP, BGP, respectively. Let $X = O_6O_1 \cap O_2O_3$, $Y = O_2O_3 \cap O_4O_5$, $Z = O_4O_5 \cap O_6O_1$, and let E, F be the points on ZX, XY, respectively, such that $GE \parallel XY$ and $GF \parallel ZX$.

By Lemma 1, we have that quadrilateral GO_1XO_2 is harmonic, which then by Lemma 2 implies that E, F, O_1 and O_2 all lie on a circle. On the other hand, we can easily check that $XE = \frac{2}{3} \cdot XO_6$, $XF = \frac{2}{3} \cdot XO_3$, which immediately implies that $O_3O_6 \parallel EF$. Therefore, points O_1, O_2, O_3, O_6 are concyclic. Analogously, we also get that O_2, O_3, O_4, O_5 are concyclic, and that O_4, O_5, O_6, O_1 are concyclic. Hence, by the radical center argument from the solution of Problem 4 (a), we conclude that the points O_1, O_2, O_3, O_4, O_5, O_6 all lie on the same circle. This completes the proof.

Solution 2. Let us now give a completely different solution that avoids projective geometry. To avoid confusion, we change the notation a bit and let M_a, M_b, M_c be the midpoints of the sides BC, CA, AB, respectively, of triangle ABC. Furthermore, let G be its centroid. The problem now states that the circumcenters A_b, C_b, C_a, B_a, B_c, A_c of triangles AM_bG, CM_bG, CM_aG, BM_aG, BM_cG, AM_cG are concyclic.

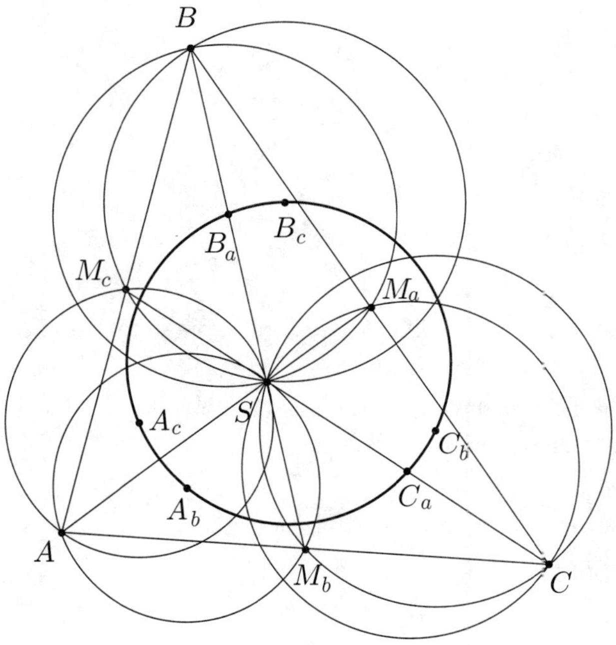

We know that the areas of these six triangles are equal. We shall further denote them by k.

Take the circumcenters B_a and B_c; they both lie on the perpendicular bisector of the segment BG. Hence, $B_aB_c \perp BG$. On the other hand, the circumcenters A_b and C_b both lie on the perpendicular bisector of the segment GM_b; hence, $A_bC_b \perp GM_b$. Since the lines supporting the segments BG and GM_b coincide, we have $B_aB_c \| A_bC_b$. Analogously, we can show that $A_cA_b \| C_aB_a$ and $C_bC_a \| B_cA_c$. Therefore, the opposite sides of the hexagon $A_bA_cB_cB_aC_aC_b$ are parallel.

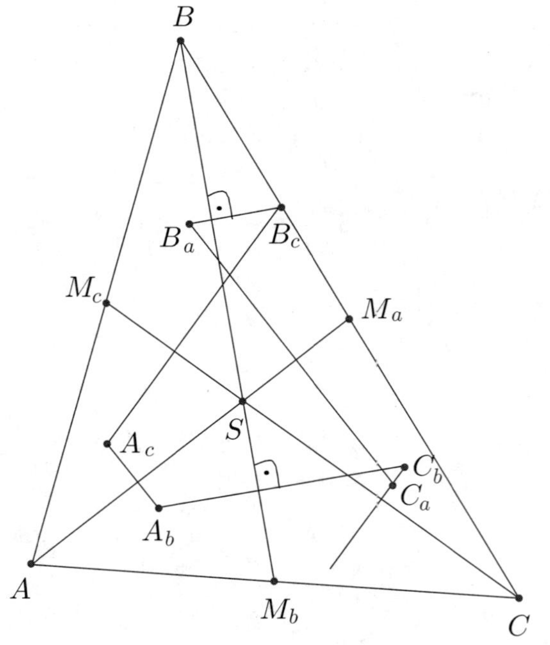

To prove the desired cyclicity we shall now make use of the following claim.

Claim. A hexagon, whose opposite sides are parallel, and whose main diagonals are of equal length, has a circumcircle.

We leave the reader the pleasure of finding the synthetic proof for this little gem. We show that

$$A_bB_a = A_cC_a = B_cC_b.$$

We will compute the length of A_cC_a by making use of the Law of Cosines in triangle A_cGC_a. For this, we must know the two other sides and the opposite angle. The side A_cG is the circumradius of triangle AM_cG; so we have

$$k = \frac{AG \cdot GM_c \cdot M_cA}{4 \cdot A_cG} = \frac{AG \cdot \frac{1}{2}CG \cdot \frac{1}{2}c}{4 \cdot A_cG} = \frac{AG \cdot CG \cdot C}{16 \cdot A_cG},$$

and hence,

$$A_cG = \frac{AG \cdot CG \cdot c}{16k}.$$

Analogously, we get

$$C_aG = \frac{AS \cdot CG \cdot a}{16k}.$$

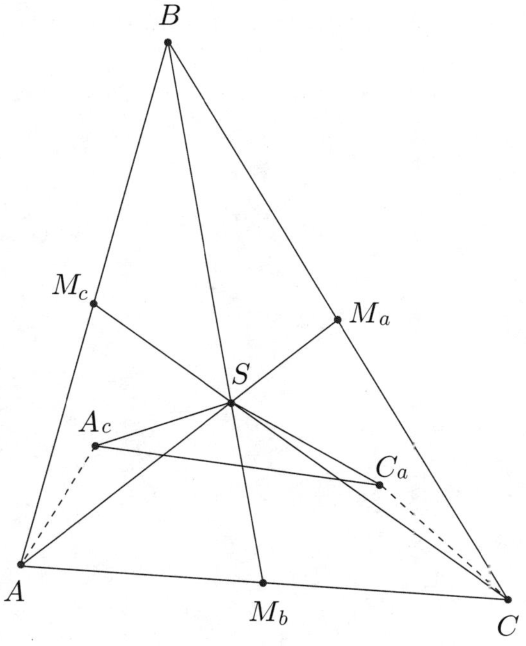

It remains to compute the magnitude of the angle A_cGC_a; in the isosceles triangle AA_cG, we have

$$\angle A_cGA = 90^\circ - \frac{1}{2}\angle AA_cG = 90^\circ - \angle AM_cG,$$

and similarly $\angle C_aGC = 90^\circ - \angle SM_aC$. Thus,

$$\begin{aligned}
\angle A_cGC_a &= \angle A_cGA + \angle AGC + \angle C_aGC \\
&= (90^\circ - \angle AM_cG) + \angle AGC + (90^\circ - \angle GM_aC) \\
&= (180^\circ - \angle AM_cG) - \angle GM_aC + \angle AGC \\
&= (180^\circ - \angle AM_cG) - \angle GM_aC + \angle AGC + (180^\circ - \angle M_cGA) \\
&= (180^\circ - \angle AM_cG) - \angle GM_aC + \angle AGC + (180^\circ - \angle GM_aC) \\
&= \angle M_cAG + (180^\circ - \angle GM_aC) \\
&= \angle BAM_a + \angle GM_aB \\
&= \angle BAM_a + \angle AM_aB \\
&= 180^\circ - B.
\end{aligned}$$

Now, we can finally apply the Law of Cosines to triangle A_cGC_a:

$$\begin{aligned}
A_cC_a^2 &= A_cG^2 + C_aG^2 - 2 \cdot A_cG \cdot C_aG \cdot \cos \angle A_cGC_a \\
&= \left(\frac{AG \cdot CG \cdot c}{16k}\right)^2 + \left(\frac{AG \cdot CG \cdot a}{16k}\right)^2 \\
&- 2 \cdot \left(\frac{AG \cdot CG \cdot c}{16k}\right) \cdot \left(\frac{AG \cdot CG \cdot a}{16k}\right) \cdot \cos(180^\circ - B) \\
&= \left(\frac{AG \cdot CG}{16k}\right)^2 \cdot \left(c^2 + a^2 - 2ca \cdot \cos(180^\circ - B)\right) \\
&= \left(\frac{AG \cdot CG}{16k}\right)^2 \cdot \left(c^2 + a^2 + 2ca \cdot \cos B\right) \\
&= \left(\frac{AG \cdot CG}{16k}\right)^2 \cdot (2 \cdot BM_b)^2 \\
&= \left(\frac{AG \cdot CG}{16k}\right)^2 \cdot \left(2 \cdot \frac{3}{2} \cdot BG\right)^2 \\
&= \left(\frac{AG \cdot CG}{16k}\right)^2 \cdot (3 \cdot BG)^2 = \left(\frac{3}{16} \cdot \frac{AG \cdot BG \cdot CG}{k}\right)^2,
\end{aligned}$$

which by noticing the symmetry in AG, BG, CG, yields that

$$A_bB_a = A_cC_a = B_cC_b.$$

According to the Claim we mentioned above, this completes the proof. ■

Problem 75. (China IMO TST 2011) Let H be the orthocenter of an acute triangle ABC with circumcircle Γ. Let P be a point on the arc AB (not containing C) of Γ, and let M be a point on the arc CA (not containing B) of Γ such that H lies on the segment PM. Let K be another point on Γ such that KM is parallel to the Simson line of P with respect to triangle ABC. Let Q be another point on Γ such that $PQ \parallel AB$. Segments AB and KQ intersect at a point J. Prove that triangle KJM is an isosceles triangle.

Solution. Let P_a, P_b, P_c, and P' be the reflections of P across BC, AC, AB, and UV, respectively, where U and V are the orthogonal projections of P onto AC and BC, respectively. By the Lemma from the proof of Problem 17, all these points lie on line $P'H$ (via a homothety with center P and ratio 2). Plus, we know that UV bisects PH and $UV \| HP_b$. Let B_1 be the reflection of H across AC, and let P_bB_1 meet the circumcircle again at K'. We then see that $\angle K'MH = 180 - \angle P_bB_1P = \angle P_bHM$, so $MK' \| P_bH \| UV$, thus $K' = K$.

Thus B_1K is the reflection of PH across AC. Hence, we get that the reflection of K with respect to AC lies on PH; therefore, if C_1 is the reflection of H across AB, the lines KC_1 , PH and AB are concurrent.

Now, let MJ meet the circumcircle of ABC again at R. By similar triangles (or by the Hexagon Ceva Lemma from the solution of Problem 105), we have that

$$QR \cdot BK \cdot AM = AQ \cdot BR \cdot MK$$

and

$$MA \cdot PC_1 \cdot BK = MK \cdot AC_1 \cdot PB.$$

Thus, we get that

$$\frac{RQ}{RB} = \frac{PC_1}{AC_1},$$

and thus $ABRC_1$ is an isosceles trapezoid. Next, let HC_1 meet UV and AB at C_2 and S, respectively, and let N be the midpoint of PH and W the projection of P onto AB. We see immediately that $HSPW$ is a parallelogram, so

$$\angle SVC_2 = \angle NC_2V = \angle PH_1 = \angle QRH_1,$$

and since $RH_1 \| AB$, we conclude that $RQ \| MK$, implying that $MQRK$ is an isosceles trapezoid. This completes the proof.

■

Problem 76. (Sammy Luo and Cosmin Pohoata, Mathematical Reflections) Let ABC be a triangle and let D, E, F be the tangency points of the incircle with BC, CA, AB, respectively. Let EF meet the circumcircle Γ of ABC at X and Y. Furthermore, let T be the second intersection of the circumcircle of DXY with the incircle. Prove that AT passes through the tangency point A' of the A-mixtilinear incircle with Γ.

Solution 1. Consider the inversion Ψ with center I and power r^2 where r is the inradius of triangle ABC. For any point X, let its inverse under Ψ be X_1. The images A_1, B_1, C_1 are on the nine-point circle (N) of triangle DEF. Points X_1, Y_1 are the intersections of (N) with the circumcircle of triangle EIF and T is the intersection of the circumcircle of triangle DX_1Y_1 with (N). Moreover, from the Lemma we proved for Problem 63, we also know that the inverse of the tangency point of the A-mixtilinear incircle with Γ tangency point is A_1', the antipode of A_1 with respect to (N). Hence, in order to prove that AT passes through A', it suffices to prove that quadrilateral $TA_1'IA_1$ cyclic.

Let the circumcircle of triangle EIF be ω, and let the circumcircle of DX_1Y_1 be ω_2. Lines DT, EF, X_1Y_1 are the pairwise radical axes of circles

ω, ω_2, and (I), so they must concur at some point Z. Similarly, the pairwise radical axes of $\omega_2, (N), (I)$ must concur; since two of these, X_1Y_1 and AT, concur at Z, the radical axis of $(N), (I)$ must also pass through Z. Let H_d, H_e, H_f be the feet of the altitudes to EF, FD, DE, respectively, in triangle DEF, and let H be its orthocenter. EFH_eH_f is cyclic with center A_1; call its circumcircle Γ_a. The pairwise radical axes of $\Gamma_a, (N), (I)$ must concur. These radical axes are H_eH_f, H_dA_1 (which is EF) and the radical axis of $(N), (I)$. The latter two pass through Z, and therefore H_eH_f must as well.

Now, notice that $D = EH_f \cap FH_e$, $H = EH_e \cap FH_f$, and $Z = EF \cap H_eH_f$. Because EFH_eH_f is cyclic, its Miquel point is the unique point that lies on DZ and is concyclic with D, E, F. Since T satisfies these properties, it must thus be the Miquel point of EFH_eH_f, so quadrilaterals TDH_eH_f, $TZEH_f$, $TZFH_e$ are also cyclic. Moreover, because EFH_eH_f is cyclic, its Miquel point T is the inverse of $EH_e \cap FH_f = H$ with respect to Γ_a, so T, H, A_1 are collinear and the line through them is perpendicular to DZ.

With N as the nine-point center of DEF, notice that A_1, A_1' and H, I are reflections of each other over N, so IA_1HA_1' is a parallelogram with center N. Then, using directed angles modulo $180°$, we write

$$\begin{aligned}
\angle A_1A_1'I &= \angle A_1'A_1H \\
&= \angle A_1'A_1H_d - \angle HA_1H_d \\
&= 90^\circ - \angle H_dA_1'A_1 - \angle TMZ \\
&= \angle MZT - \angle H_dH_fA_1 \\
&= \angle FZD - (\angle EH_fA_1 - \angle EH_fH_d) \\
&= \angle FZD - (\angle A_1EH_f - \angle DFE) \\
&= \angle FZD - \angle FED + \angle DFE.
\end{aligned}$$

But we also have that

$$\begin{aligned}
\angle A_1TI &= 90^\circ - \angle ITD \\
&= \angle H_eFT \\
&= \angle H_eZT \\
&= \angle MZT - \angle FZH_e \\
&= \angle MZT - (\angle FH_eZ + \angle ZFH_e) \\
&= \angle MZT - \angle FEH_f - \angle EFD \\
&= \angle FZD - \angle FED + \angle DFE.
\end{aligned}$$

Thus, $\angle A_1A_1'I = \angle A_1TI$, and therefore A_1, A_1', T, I are indeed concyclic, as desired. This completes the proof.

Solution 2. This approach, while computational, is very instructive, since it displays an ingenious interplay between the mechanical algebra and the geometric ideas that lie behind it.

Denote by O_a the center of the mixtilinear incircle. We saw before in the proof of Problem 36 that the mixtilinear inradius can be computed to be $\rho_a = \frac{r}{\cos^2 \frac{A}{2}}$; or the exact trilinear coordinates of O_a are

$$O_a \equiv \left(\rho_a \frac{1 + \cos A - \cos B - \cos C}{2}, \rho_a, \rho_a\right).$$

Consider now the second intersection Z', other than A, of line $c(s-b)\beta = b(s-c)\gamma$ with the circumcircle of ABC. Since the equation of the latter is $a\beta\gamma + b\gamma\alpha + c\alpha\beta = 0$, the coordinates of Z' satisfy $c(s-b)\beta = b(s-c)\gamma = -bc\alpha$, or in non-exact trilinear coordinates, after some algebra we find

$$Z' \equiv \left(-\frac{r}{2R}, 1 - \cos B, 1 - \cos C\right).$$

We can furthermore verify after some algebra that

$$(1 + \cos A - \cos B - \cos C, 2, 2), \ (\cos A, \cos B, \cos C),$$

$$\left(-\frac{r}{2R}, 1 - \cos B, 1 - \cos C\right)$$

are linearly dependent as vectors, or Z' is on the circumcircle, line OO_a, and the sector determined by rays AB, AC, i.e., $Z' = A'$ is the point where the mixtilinear incircle touches the circumcircle. Indeed, this linear dependence is equivalent to $\cos A + \cos B + \cos C = 1 + \frac{r}{R}$, true by Carnot's theorem. The problem then reduces to proving that the intersection Z of line $c(s-b)\beta = b(s-c)\gamma$ and the incircle, furthest from A, lies on the circumcircle of DXY, where clearly Z is the result of scaling Z' with respect to A and a scaling factor $\cos^2 \frac{A}{2}$ (ie, the same scaling that transforms the A-mixtilinear incircle into the incircle).

Now, using that $c(s-c) = 4Rr\cos^2 \frac{C}{2}$, we may rewrite the equation for the incircle in trilinear coordinates as

$$a^2(s-a)^2\alpha^2 + b^2(s-b)^2\beta^2 + c^2(s-c)^2\gamma^2 =$$

$$= 2bc(s-b)(s-c)\beta\gamma + 2ca(s-c)(s-a)\gamma\alpha + 2ab(s-a)(s-b)\alpha\beta.$$

After some algebra, the intersection with line $c(s-b)\beta = b(s-c)\gamma$ closest to side BC (therefore further from A) satisfies

$$abc(s-a)\alpha = (b-c)^2 c(s-b)\beta = (b-c)^2 b(s-c)\gamma.$$

For α, β, γ to be exact trilinear coordinates, we need $a\alpha + b\beta + c\gamma = 2[ABC]$, or equivalently, in exact trilinear coordinates

$$Z \equiv \left(\frac{(b-c)^2(s-b)(s-c)}{2R\,(a(s-a)+(b-c)^2)}, \frac{ab(s-c)(s-a)}{2R\,(a(s-a)+(b-c)^2)}, \frac{ca(s-a)(s-b)}{2R\,(a(s-a)+(b-c)^2)} \right).$$

It now suffices to show that this point is concyclic with D, X, Y.

Now, given four points with exact trilinear coordinates (α, β, γ) with respect to a reference triangle ABC, it is well known that they are concyclic if and only if the four 4-vectors of the form $(a\beta\gamma + b\gamma\alpha + c\alpha\gamma, \alpha, \beta, \gamma)$ are collinear. Denote by $\vec{u}, \vec{v}$ the respective 4-vectors for D, Z. A linear combination with not both zero coefficients ρ, κ such that $\rho\vec{u} + \kappa\vec{v} \equiv (0, \alpha', \beta', \gamma')$. Clearly, $a\beta\gamma + b\gamma\alpha + c\alpha\gamma = 0$ for X, Y because they are both on the circumcircle, and since they are also both on line EF, it suffices to show that $(\alpha', \beta', \gamma')$ satisfies the equation for line EF.

Note that, in exact trilinear coordinates,

$$D \equiv (0, (s-c)\sin C, (s-b)\sin B)\,,$$

and similarly for E, F. It follows that the 4-vector for D is

$$\vec{u} \equiv (a(s-b)(s-c)\sin B \sin C, 0, (s-c)\sin C, (s-b)\sin B)\,,$$

while the equation for line EF is

$$EF \equiv a(s-a)\alpha = b(s-b)\beta + c(s-c)\gamma.$$

The 4-vector for Z is found from previous results after a few calculations and using Heron's formula:

$$\vec{v} \equiv \left(\frac{arS^2}{KR}, \frac{(b-c)^2(s-b)(s-c)}{2KR}, \frac{ab(s-c)(s-a)}{2KR}, \frac{ca(s-a)(s-b)}{2KR} \right),$$

where $K = a(s-a) + (b-c)^2$.
It follows that $\kappa = KRbc$, $\rho = -4R^2rs(s-a) = -abcR(s-a)$ yields the desired linear combination, for which

$$(\alpha', \beta', \gamma') \equiv \left(K' \frac{(b-c)(s-b)(s-c)}{a}, (s-c)(s-a), -(s-a)(s-b) \right)$$

where $K' = \frac{abc(b-c)}{2}$. It then suffices to show that

$$(s-a)(b-c)(s-b)(s-c) = b(s-b)(s-c)(s-a) - c(s-c)(s-a)(s-b),$$

clearly true. The conclusion follows.

■

Problem 77. (Titu Andreescu and Cosmin Pohoata, USAMO Shortlist 2014) Let ABC be a triangle with incircle γ and circumcircle Γ. Let Ω be the circle tangent to the rays AB, AC and to Γ externally, and let A' be the tangency point of Ω with Γ. Furthermore, draw the tangents from A' with respect to incircle γ and let B', C' be their other intersections with Γ. If X denotes the tangency point of the chord $B'C'$ with γ, prove that the circumcircle of triangle BXC is tangent to γ.

Solution. First, just to make sense of the statement, recall the famous Poncelet porism which assures the tangency of $B'C'$ with the incircle γ. We state a reformulation of the problem which is easier to work with.

Reformulation. Suppose X is instead defined to be the tangency point of the incircle with the circle passing through vertices B and C which is tangent to the incircle. Then, take the tangent ℓ at X with respect to γ and denote its intersection points with the circumcircle by B' and C'. Furthermore, draw the tangents from B' and C' to γ that are different from ℓ. Again, Poncelet's porism assures that these tangents meet at a point A' lying on the circumcircle Γ. Then, if D is the tangency point of the γ with BC, we have to prove that $\angle DAB = \angle A'AC$, or in other words, the lines AD and AA' are isogonal with respect to $\angle A$.

The equivalence between the two versions is given by the following claim about the configuration.

Claim 1. In the notations from the original formulations, we have that $\angle DAB = \angle A'AC$.

This is a statement about mixtilinear excircles. The line AD is a Gergonne cevian, whereas AA' passes through the insimilicenter of γ and Γ, which is the isogonal conjugate of the Gergonne point of triangle ABC.

We now turn to the reformulation.

Proof of Reformulation. We begin with a very important Lemma that lies at the heart of the problem.

Lemma. Let ABC and $A'B'C'$ be two triangles that have the same circumcircle $\Gamma(O, R)$ and the same incircle $\gamma(I, r)$. Let D, E, F, D', E', F' be the tangency points of γ with the chords BC, CA, AB, $B'C'$, $C'A'$, $A'B'$, respectively. Then, the line passing through the midpoints of segments EF and $E'F'$ is parallel to DD'.

Proof of Lemma. We assume without loss of generality that points A', B', C' lie on the small arcs CA, AB, BC. Define the intersections $P = BC \cap B'C'$, $Q = AB \cap B'C'$, $R = AB \cap B'C'$, $S = AC \cap A'B'$.

Consider the inversion Ψ with center I and power r^2, where r is the inradius of triangles ABC and $A'B'C'$. We know from the proof of Problem 40 that this inversion takes the vertices A and A' to the midpoints of segments EF and $E'F'$, and thus Ψ maps the line through the midpoints of EF, $E'F'$ into the circumcircle of triangle IAA'. It is therefore sufficient to show that the tangent of circle (IAA') at vertex I is parallel to DD'. In particular, we just have to show that P passes through the circumcenter of IAA'.

Note that

$$\begin{aligned}\angle A'C'B' &= \angle AC'A' + \angle AC'B' \\ &= \angle AC'A' + \angle AA'B' \\ &= \angle ABC - \angle CAA' + \angle AA'B';\end{aligned}$$

hence

$$\frac{1}{2}\angle A'C'B' - \angle AA'B' = \frac{1}{2}\angle ABC - \frac{1}{2}\angle CAA' - \frac{1}{2}\angle AA'B',$$

which gives

$$\begin{aligned}90^\circ - \frac{1}{2}\angle B'A'C' &= 90^\circ - \angle IA'A \\ &= \frac{1}{2}\angle ABC + \frac{1}{2}\angle A'B'C' - \frac{1}{2}\angle CAA' - \frac{1}{2}\angle AA'B'.\end{aligned}$$

We label this relation as (1) for reference purposes. Now, let $L = AI \cap BC$ and suppose L is between C and P. Using that I is the common excenter of triangles BPQ, $B'QR$ and ARS, we get

$$\begin{aligned}\angle PIA &= \angle PIQ + \angle QIR + \angle RIA \\ &= \left(90^\circ - \frac{1}{2}\angle ABC\right) + \left(90^\circ - \frac{1}{2}\angle A'B'C'\right) + \frac{1}{2}\angle ASR \\ &= 180^\circ - \frac{1}{2}\angle ABC - \frac{1}{2}\angle A'B'C' + \frac{1}{2}\angle CAA' + \frac{1}{2}\angle AA'B'.\end{aligned}$$

Therefore, by (1),

$$\angle PIL = 180^\circ - \angle PIA = 90^\circ - \angle IAA',$$

which proves that PI passes through the circumcenter of triangle IAA', as claimed. This proves the Lemma.

□

Now, let us return to the proof of our Reformulation. Let E, F, D', E', F' be the tangency points of the incircle γ with the chords CA, AB, $B'C'$, $C'A'$, $A'B'$, respectively. Note that the previous point X was replaced by D' for convenience of notation. Furthermore, let $T_1 = EF \cap BC$, $T_1' = E'F' \cap B'C'$, $P = BC \cap B'C'$, and $T = E'F' \cap EF$. We consider polars only with respect to γ.

As already established, we know from Poncelet's porism that A' lies on the circumcircle Γ. We also know that the polars of A, F, D, A', E are $EF, AB, BC, E'F', AC$, respectively. Consequently, AD is the polar of T_1 and AA' is the polar of T. Hence, $IT_1 \perp AD$ and $IT \perp AA'$; thus, in order to show that $\angle DAB = \angle A'AC$ or equivalently $\angle IAD = \angle IAA'$, it would suffice to show that $\angle FIT_1 = \angle TIE$, or equivalently $TI = T_1I$.

On the other hand, the quadruple (T_1, B, D, C) is a harmonic division and P lies on BC so that $PB \cdot PC = (PD')^2 = PD^2$ (by Power of a Point); hence P is the midpoint of segment T_1D. Using an analogous argument we also get that P is the midpoint of $T_1'D'$, implying that $DD'T_1'T_1$ is a rectangle. Because I lies on the perpendicular bisector of DD', it means that $T_1I = T_1'I$. So all that is left to do is to prove that I is the circumcenter of triangle TT_1T_1'.

Here is where the Lemma comes in. Let M_1, M_1' be the midpoints of segments T_1T and TT_1', respectively. By the Lemma, we have that M_1M_1' is parallel to DD', so M_1M_1' is parallel to T_1T_1'. However, the incenter I is on the perpendicular bisector of T_1T_1' as argued above. Thus, by Carnot's Theorem applied for triangle TT_1T_1', we have that

$$(M_1T^2 - M_1T_1^2) + 0 + (M_1'T_1'^2 - M_1'T^2) = 0;$$

hence $M_1T^2 - M_1T_1{}^2 = M_1'T^2 - M_1'T_1'^2$. Combining this with the fact that

$$\frac{M_1T}{M_1T_1} = \frac{M_1'T}{M_1'T_1'}$$

(which comes from $M_1M_1' \| T_1T_1'$), we get that M_1, M_1' are the midpoints of TT_1 and TT_1', from which the result follows. This completes the proof.

■

Problem 78. (AOPS Forum) In triangle ABC, let D, E, F be the feet of the altitudes from A, B, and C, respectively. Let H be the orthocenter of triangle ABC and let I_1, I_2, I_3 be the incenters of triangles EHF, FHD, and DHE, respectively. Prove that the lines AI_1, BI_2, CI_3 are concurrent.

Solution. The key idea is to use Jacobi's theorem from the proof of Problem 22. However, there's no immediate way to apply it here, so we we will have to design our own setting.

Construct the triangle XBC outside of triangle ABC such that triangles XBC and HEF are similar. Similarly, define Y and Z such that triangles YCA and HFD are similar and such that ZAB and HDE are similar. Notice that triangle ABC is similar to triangle AEF (since $BCEF$ is cyclic); thus, quadrilaterals $AEHF$ and $ABXC$ are similar. In particular, if I_1' denotes the incenter of triangle XBC, it follows that angles $\angle I_1AB = \angle I_1AF = \angle I_1'AC$. Analogously, if I_2' denotes the incenter of YCA and I_3' the incenter of ZAB, we get that $\angle I_2BC = \angle I_2'BA$ and $\angle I_3CA = \angle I_3'CB$. Consequently, lines AI_1, BI_2, CI_3 are concurrent if and only if lines AI_1', BI_2', CI_3' are concurrent (since the latter are simply the reflections of the three original lines across the corresponding internal angle bisectors).

On the other hand, $\angle I_1'BC = \frac{1}{2}\angle XBC = \frac{1}{2}\angle HEF$ and $\angle I_3'BA = \frac{1}{2}\angle ZBA = \frac{1}{2}\angle HED$. But $\angle HEF = \angle HED$, since BH is the internal angle bisector of angle $\angle D$ of DEF. Hence, $\angle I_1'BC = \angle I_3'BA$, and, similarly, it follows that $\angle I_1'CB = \angle I_2'CA$ and $\angle I_2'AC = \angle I_3'AB$, as well. Therefore, lines AI_1', BI_2', CI_3' are concurrent, by Jacobi's theorem. This completes the proof. ■

Problem 79. (Eric Daneels, Forum Geometricorum) Let ABC be a triangle with incircle Γ and let D, E, F be the tangency points of Γ with BC, CA, AB, respectively. Let M, N, P be the midpoints of BC, CA, AB, and let X, Y, Z be points on the lines AI, BI, CI, respectively. Prove that the lines XD, YE, ZF are concurrent if and only if the lines XM, YN, ZP are concurrent.

Solution. Let τ_A, τ_B, τ_C, ℓ_A, ℓ_B, ℓ_C denote the lines EF, FD, DE, NP, PM, and MN, respectively. Also, let $\delta(U, \ell)$ stand for the (unsigned) distance from the point U to the line ℓ.

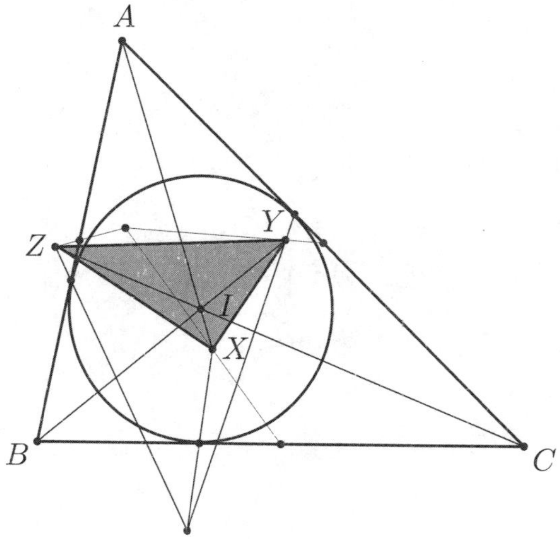

Recall the Lemma from Problem 32 that the polar of one vertex with respect to the incircle, the midline corresponding to a second vertex, and the internal angle bisector issuing from the third are concurrent. This is what lies at the heart of this magnificent result. In particular, the lines ED, PM, AI concur, as do the lines FD, NM, AI. Consequently, any point X on the line AI satisfies

$$\frac{\delta(X,\tau_C)}{\delta(X,\ell_B)} = \frac{\delta(I,\tau_C)}{\delta(I,\ell_B)} \quad \text{and} \quad \frac{\delta(X,\tau_B)}{\delta(X,\ell_C)} = \frac{\delta(I,\tau_B)}{\delta(I,\ell_C)},$$

which gives

$$\frac{\delta(X,\tau_C)}{\delta(X,\tau_B)} \cdot \frac{\delta(X,\ell_C)}{\delta(X,\ell_B)} = \frac{\delta(I,\tau_C)}{\delta(I,\tau_B)} \cdot \frac{\delta(I,\ell_C)}{\delta(I,\ell_B)}.$$

Analogously, any points Y and Z on the lines BI and CI, respectively, satisfy

$$\frac{\delta(Y,\tau_A)}{\delta(Y,\tau_C)} \cdot \frac{\delta(Y,\ell_A)}{\delta(Y,\ell_C)} = \frac{\delta(I,\tau_A)}{\delta(I,\tau_C)} \cdot \frac{\delta(I,\ell_A)}{\delta(I,\ell_C)}$$

and

$$\frac{\delta(Z,\tau_B)}{\delta(Z,\tau_A)} \cdot \frac{\delta(Z,\ell_B)}{\delta(Z,\ell_A)} = \frac{\delta(I,\tau_B)}{\delta(I,\tau_A)} \cdot \frac{\delta(I,\ell_B)}{\delta(I,\ell_A)}.$$

Hence, by multiplying these all together, it follows that

$$\begin{aligned} &\left[\frac{\delta(X,\tau_C)}{\delta(X,\tau_B)} \cdot \frac{\delta(Y,\tau_A)}{\delta(Y,\tau_C)} \cdot \frac{\delta(Z,\tau_B)}{\delta(Z,\tau_A)}\right] \cdot \left[\frac{\delta(X,\ell_C)}{\delta(X,\ell_B)} \cdot \frac{\delta(Y,\ell_A)}{\delta(Y,\ell_C)} \cdot \frac{\delta(Z,\ell_B)}{\delta(Z,\ell_A)}\right] \\ = &\left[\frac{\delta(I,\tau_C)}{\delta(I,\tau_B)} \cdot \frac{\delta(I,\tau_A)}{\delta(I,\tau_C)} \cdot \frac{\delta(I,\tau_B)}{\delta(I,\tau_A)}\right] \cdot \left[\frac{\delta(I,\ell_C)}{\delta(I,\ell_B)} \cdot \frac{\delta(I,\ell_A)}{\delta(I,\ell_C)} \cdot \frac{\delta(I,\ell_B)}{\delta(I,\ell_A)}\right] \\ = &\ 1. \end{aligned}$$

In particular,

$$\left[\frac{\delta(X,\tau_C)}{\delta(X,\tau_B)} \cdot \frac{\delta(Y,\tau_A)}{\delta(Y,\tau_C)} \cdot \frac{\delta(Z,\tau_B)}{\delta(Z,\tau_A)}\right] = 1$$

if and only if

$$\left[\frac{\delta(X,\tau_C)}{\delta(X,\tau_B)} \cdot \frac{\delta(Y,\tau_A)}{\delta(Y,\tau_C)} \cdot \frac{\delta(Z,\tau_B)}{\delta(Z,\tau_A)}\right] = 1.$$

Thus, by writing out Ceva's theorem in triangles DEF and MNP, we conclude that the lines DX, EY, FZ concur if and only if MX, NY, PZ concur. This completes the proof.

■

Problem 80. (Dan Schwarz, Stars of Mathematics 2008) Consider a convex quadrilateral and the incircles of the triangles determined by one of its diagonals. Prove that the tangency points of the incircles with the diagonal are symmetric with respect to the midpoint of the diagonal if and only if the diagonals and the line through the incenters are concurrent.

Solution. This is a problem that involves two classical theorems that are applied in non-classical ways (which makes it a good contest problem).

Pitot's Ex-Theorem. Let $ABCD$ be a convex quadrilateral such that $AB+BC = CD+DA$. Then, $ABCD$ has an inscribed circle that is tangent to all the sides of the quadrilateral.

Monge-d'Alembert's In-Theorem. Two of the insimilicenters determined by three distinct circles, all lying in the same plane, are collinear with the exsimilicenter of the last pair of circles.

We omit the proofs of these two variants since they are identical to the proofs for the standard versions of the theorems. Returning to the problem, suppose that $ABCD$ is the convex quadrilateral and that (I) and (J) are incircles of triangles DAB and BCD, respectively. Furthermore, suppose that (I) and (J) are tangent to BD at points U and V, respectively. The internal homothety center of (I) and (J) is $H = IJ \cap BD$. If $\angle DAB = \angle BCD$, either condition $BU = DV$ or $H = IJ \cap BD \cap AC$ makes $ABCD$ a parallelogram and the proposition becomes trivial. Therefore, without loss of generality, let $\angle DAB < \angle BCD$ and $P \equiv AB \cap CD$.

Assume that $BU = DV$. Then

$$AB + BD - DA = CD + DB - BC$$

or $AB+BC = CD+DA$, which, by the Pitot variant, means that the convex $ABCD$ is tangent to an excircle (E) in the (smaller) angle $\angle DAB$. Furthermore, note that A is the external homothety center of (I) and (E) and C is the internal homothety center of (J) and (E), so, by the variant of Monge-d'Alembert's theorem, the internal homothety center H of $(I),(J)$ is on A; therefore $H = IJ \cap BD \cap AC$.

Conversely, suppose $H = IJ \cap BD \cap AC$ and let (E) be the A-excircle of triangle DAP. The internal homothety center of (J) and (E) is on their common internal tangent $CD \equiv PD$. Point A is the external homothety center of (I) and (E) and H is the internal homothety center of (I) and (J). The variant of the Monge-d'Alembert theorem we stated thus yields that the internal homothety center of (J) and (E) is on AC; in particular, it is identical to $C = CD \cap AC$. But then the other tangent BC to (J) through C is also tangent to (E), and so the convex quadrilateral $ABCD$ is tangent to the excircle (E) in the angle $\angle DAB$. Consequently, $AB + BC = CD + DA$ or $AB + BD - DA = CD + DB - BC$, which means that $BU = DV$. This completes the proof.

■

Problem 81. (Luis Gonzalez, AOPS Forum) A circle $\mathcal{K}$ passes through the vertices B, C of triangle ABC and another circle ω touches AB, AC, $\mathcal{K}$ at P, Q, T, respectively. If M is the midpoint of the arc BC of $\mathcal{K}$ that contains the point T, prove that the lines BC, PQ, and MT are concurrent.

Solution. The result that lies at the heart of the solution is the following very useful formula for the length of the common external tangent.

Casey's Chord Theorem. Let circles Γ_1, Γ_2, having radii r_1 and r_2, respectively be internally/externally tangent to a circle Γ, having radius R, at points A and B, respectively. Then, the length δ_{12} of the common external tangent between Γ_1 and Γ_2 is given by

$$\delta_{12} = \frac{AB}{R}\sqrt{(R \pm r_1)(R \pm r_2)},$$

where the choice of signs depends on whether Γ_1, Γ_2 are internally or externally tangent to Γ.

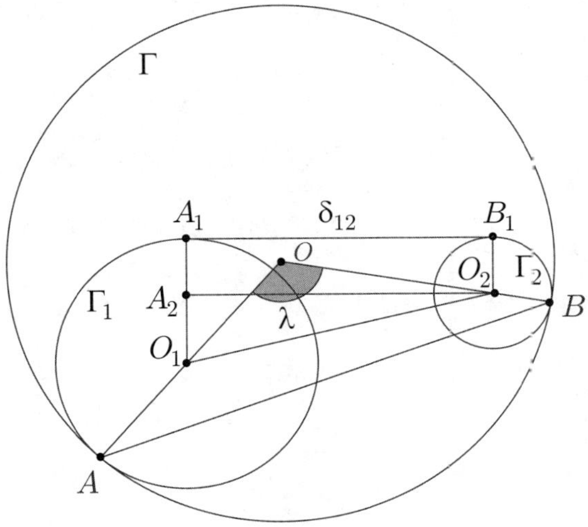

Proof. Without loss of generality assume that $r_1 \geq r_2$ and suppose that Γ_1 and Γ_2 are internally tangent to Γ. The other cases can be treated analogously. A common external tangent between Γ_1 and Γ_2 touches Γ_1, Γ_2 at A_1, B_1, respectively, and A_2 is the orthogonal projection of O_2 onto O_1A_1. By the Pythagorean theorem for $O_1O_2A_2$, we obtain

$$\delta_{12}^2 = (A_1B_1)^2 = (O_1O_2)^2 - (r_1 - r_2)^2.$$

Let $\angle O_1OO_2 = \lambda$. By the Law of Cosines in triangle OO_1O_2, we get

$$(O_1O_2)^2 = (R - r_1)^2 + (R - r_2)^2 - 2(R - r_1)(R - r_2)\cos\lambda$$

By the Law of Cosines in the isosceles triangle OAB, we get

$$AB^2 = 2R^2(1 - \cos\lambda).$$

Eliminating $\cos\lambda$ and O_1O_2 from the three previous expressions yields

$$\delta_{12}^2 = (R - r_1)^2 + (R - r_2)^2 - (r_1 - r_2)^2 - 2(R - r_1)(R - r_2)\left(1 - \frac{AB^2}{2R^2}\right).$$

Subsequent simplications give

$$\delta_{12} = \frac{AB}{R}\sqrt{(R - r_1)(R - r_2)}$$

Analogously, if Γ_1, Γ_2 are externally tangent to Γ, then we will get

$$\delta_{12} = \frac{AB}{R}\sqrt{(R + r_1)(R + r_2)}.$$

If Γ_1 is externally tangent to Γ and Γ_2 is internally tangent to Γ, then a similar reasoning gives that the length of common internal tangent between Γ_1 and Γ_2 is given by

$$\delta_{12} = \frac{AB}{R}\sqrt{(R + r_1)(R - r_2)}.$$

This completes the proof of the Casey Chord Theorem.

Returning to the problem, let R and ϱ be the radii of $\mathcal{K}$ and ω, respectively. By Casey's Chord Theorem for ω, (B) and ω, (C), where (B) and (C) are the degenerate circles B and C, which are (trivially) tangent to $\mathcal{K}$, we get

$$TC^2 = \frac{CQ^2 \cdot R^2}{(R - \varrho)(R - 0)} = \frac{CQ^2 \cdot R}{R - \varrho}, \quad TB^2 = \frac{BP^2 \cdot R^2}{(R - \varrho)(R - 0)} = \frac{BP^2 \cdot R}{R - \varrho},$$

which yields

$$\frac{TB}{TC} = \frac{BP}{CQ}.$$

Now, let PQ intersect BC at U. By Menelaus' theorem in triangle ABC, applied for the collinear points U, P, Q, we have that

$$\frac{UB}{UC} = \frac{BP}{AP} \cdot \frac{AQ}{CQ} = \frac{BP}{CQ} = \frac{TB}{TC};$$

thus, by the Angle Bisector theorem, U is the foot of the T-external bisector TM of BTC; thus the point has to lie on the line MT. This proves the problem.

■

Problem 82. (Cosmin Pohoata, Mathematics Magazine) Let $ABCD$ be a quadrilateral with an inscribed circle Γ and let γ be an arbitrary line tangent to its incircle. Let A', B', C', D' be the orthogonal projections of A, B, C, D onto γ, respectively. Prove that

$$\frac{AA' \cdot CC'}{BB' \cdot DD'} = \frac{AO \cdot CO}{BO \cdot DO}.$$

Solution. We will start with a simple preliminary result, as usual.

Lemma. Let $ABCD$ be a cyclic quadrilateral and let P be a point on its circumscribed circle. Let X, Y, Z, W the projections of P onto the sides AD, AB, BC, and CD, respectively. Then $PX \cdot PZ = PY \cdot PW$.

Proof. Without loss of generality, suppose that P lies on the arc AB that does not contain C and D. In this case,

$$\frac{PX}{PY} = \frac{\sin PAX}{\sin PAB} \text{ and } \frac{PZ}{PW} = \frac{\sin PCB}{\sin PCD}.$$

But notice that $\angle PAB = \angle PCD$ and $\angle PAX = \angle PCD$. Therefore, combining the two relations, we get $PX \cdot PZ = PY \cdot PW$.

Returning to our problem, let M, N, L, Q be the tangency points of the incircle (O) of $ABCD$ with the sides DA, AB, CB, CD, respectively. Let X, Y, Z, W be the orthogonal projections of P on MN, NL, LQ, and QM, respectively. Also, denote by P the tangency point of γ with (O).

Notice that γ is the polar of P with respect to the circle (O) and that MN, NL, LQ, and QM are the polars of A, B, C, D, respectively. From Salmon's theorem, we have that

$$\frac{AA'}{AO} = \frac{PX}{R}, \quad \frac{CC'}{CO} = \frac{PZ}{R}$$

and

$$\frac{BB'}{BO} = \frac{PY}{R}, \quad \frac{DD'}{DO} = \frac{PW}{R},$$

where R is the inradius of quadrilateral $ABCD$. However, by the Lemma we just proved, applied in the cyclic quadrilateral $MNLQ$, we know that $PX \cdot PZ = PY \cdot PW$, whence

$$\frac{AA' \cdot CC'}{BB' \cdot DD'} = \frac{AO \cdot CO}{BO \cdot DO},$$

which completes the proof.

∎

Problem 83. (Peter Woo, Crux Mathematicorum) Let $ABCD$ be a quadrilateral with an inscribed circle and let P be the intersection of its diagonals. Prove that the incenters of triangles PAB, PBC, PCD, PDA are concyclic.

Solution. We begin with a truly beautiful lemma.

Lemma. Let ABC be a triangle with incenter I and let P be a point on the side BC. Let D be the tangency point of the incircle of ABC with BC and let I_1 and I_2 be the incenters of triangles APB and APC. Prove that the points P, D, I_1, I_2 are concyclic.

Proof. Let X and Y be the feet of the perpendiculars from I_1, I_2 to BC, respectively. Note that $\angle I_1PI_2 = 90°$, so triangles I_1XP and PYI_2 are similar. Furthermore, we note that

$$\begin{aligned} YD &= PY + BP - BD \\ &= \frac{1}{2}(AP + PC - AC) + BP - \frac{1}{2}(AB + BC - AC) \\ &= \frac{1}{2}(AP + BP - AB) \\ &= PX. \end{aligned}$$

In particular, this also means that $XD = PY$; it thus follows that

$$\frac{I_1X}{XD} = \frac{I_1X}{PY} = \frac{XP}{YI_2} = \frac{DY}{YI_2}.$$

This implies that triangles I_1XD and DYI_2 are similar, and so $\angle I_1DI_2 = 90° = \angle I_1PI_2$. This gives that P, D, I_1, I_2 lie on the same circle, as desired. □

Returning to the problem, let I_1, I_2, I_3, I_4 be the incenters of triangles PAB, PBC, PCD, PDA, respectively, and let X, Y, Z, T be the incenters of triangles ABD, ABC, BCD, CDA, respectively. Furthermore, let O be the incenter of $ABCD$.

Since lines AX, CY, BZ are concurrent at O, we can say that triangles ACB and XYZ are perspective; thus, Desargues' theorem implies that the intersections $AY \cap BX$, $BZ \cap CY$, and $XZ \cap AC$ are collinear, or, equivalently, the lines I_1I_2, AC and XZ are concurrent. Analogously, we get that I_3I_4, AC and XZ are concurrent; hence, it follows that the intersection T of I_1I_2 and I_3I_4 lies on the line AC.

Let L and L' be the feet of the perpendiculars from Y and Z to the line AC. Note that

$$CL = \frac{AC + BC - AB}{2} \quad \text{and} \quad CL' = \frac{AC + CD - AD}{2}.$$

However, $AB + CD = AD + BC$; so, $CL = CL'$, or, in other words, $L \equiv L'$. This is where the magic happens. By the Lemma we just proved, quadrilaterals I_1LPI_2 and I_3LPI_3 are cyclic, and so, the Power of Point theorem, gives

$$TI_1 \cdot TI_2 = TL \cdot TP = TI_3 \cdot TI_4.$$

This allows us to conclude that I_1, I_2, I_3, I_4 lie on the same circle, as claimed. ∎

Comment. The converse of the result is also true. Namely, if $ABCD$ is a convex quadrilateral with diagonals intersecting at P and so that the incenters of triangles PAB, PBC, PCD, PDA are concyclic, then $ABCD$ admits an inscribed circle. We leave the reader the non-trivial task of reversing the above argument.

Problem 84. (Tashio Seimiya, Crux Mathematicorum) Let P be an arbitrary interior point of an equilateral triangle ABC. Prove that

$$|\angle PAB - \angle PAC| \geq |\angle PBC - \angle PCB|.$$

Solution. Let M be the midpoint of BC. Then, AM is the perpendicular bisector of BC, and if P lies on it, the given relation holds with both sides being zero. We may therefore assume, without loss of generality, that P is an interior point of triangle ABM. We then have $\angle PAB < \angle PAC$ and $\angle PBC > \angle PCB$, so that $|\angle PAB - \angle PAC| = \angle PAC - \angle PAB$ and $|\angle PBC - \angle PCB| = \angle PBC - \angle PCB$. Thus the relation we wish to prove reduces to

$$\angle PAC - \angle PBC > \angle PBC - \angle PCB.$$

Let Q be the reflection of P in the line AM; then $\angle PAB = \angle QAC$ and $\angle PCB = \angle QBC$. Thus,

$$\angle PAC - \angle PAB = \angle PAC - \angle QAC = \angle PAQ$$

and

$$\angle PBC - \angle PCB = \angle PBC - \angle QBC = \angle PBQ,$$

so that the inequality to prove now becomes $\angle PAQ > \angle PBQ$.

Since $PQ \perp AM$ and $AM \perp BC$, the lines PQ and BC are parallel. Let PQ meet AB and AC at R and S, respectively, and let T be the reflection of B in RS. Then, $\angle PTQ = \angle PBQ$. Since AM is the perpendicular bisector of both PQ and RS, the circumcenters of triangles APQ and ARS lie on AM, so that the circumcircle of ABC is tangent at A to the circumcircles of both

ARS and APQ, which we denote by Γ and Γ', respectively. Note that Γ' is contained in Γ. Because

$$\angle TRA = \angle TRQ - \angle ARQ = \angle BRQ - \angle ARQ = 120^\circ - 60^\circ = 60^\circ,$$

while $\angle ASR = 60^\circ$, it follows that $\angle TRA = \angle ASR$. Hence RT is tangent to Γ so that T is a point outside Γ, and consequently, T is a point outside Γ'. Since A and T are on the same side of PQ, we have $\angle PAQ > \angle PTQ = \angle PBQ$. This completes the proof.

Comment. The above argument extends to triangles for which $\angle B = \angle C \leq 60^\circ$; that is $|\angle PAB - \angle PAC| \geq |\angle PBC - \angle PCB|$ also for isosceles triangles when $\angle A \geq 60^\circ$. On the other hand, when $\angle B = \angle C > 60^\circ$ one can find positions for P where the given inequality fails to hold.

■

Problem 85. (AOPS Forum) Let ρ be the inscribed circle of ABC with incenter I and let D, E and F be the points of tangency of ρ with BC, CA, and AB respectively. Denote by M the second intersection of ρ with AD, and let N be the second intersection of the circumcircle of triangle CDM with DF. Furthermore, let G be the intersection of CN and AB. Prove that $CD = 3FG$.

Solution. Let X be the intersection of EF with CG and let T be the intersection of EF with BC. The quadruplet (T, B, D, C) is a harmonic division, so the pencil (FT, FB, FD, DC) is harmonic. Hence, by intersecting it with the line CG, we get that (X, G, N, C) is a harmonic division.

By Menelaus' theorem in triangle BCG, applied for the collinear points D, N and F, we see immediately that $CD = 3FG$ happens if and only if $CN = 3NG$. But we know from the harmonic division above that

$$\frac{NC}{NG} = \frac{XC}{XG};$$

thus, it is enough to prove that N is the midpoint of segment CX.

We proceed as follows. Note that $\angle MEX = \angle MDF = \angle MCX$, therefore quadrilateral $MECX$ is cyclic. This implies that $\angle MXC = \angle MEA = \angle ADE$ and $\angle MCX = \angle ADF$. Furthermore, $\angle CMN = \angle FDB$ and

$$\begin{aligned}\angle XMN &= \angle XMC - \angle CMN \\ &= \angle CEF - \angle FDB \\ &= \angle EDC.\end{aligned}$$

By plugging the angle equalities obtained above into the Ratio Lemma identity

$$\frac{NX}{NC} = \frac{MX}{MC} \cdot \frac{\sin \angle XMN}{\sin \angle CMN} = \frac{\sin \angle MCX}{\sin \angle MXC} \cdot \frac{\sin \angle XMN}{\sin \angle CMN},$$

it follows that

$$NC = NX \text{ if and only if } \frac{\sin \angle FDA}{\sin \angle EDA} = \frac{\sin \angle BDF}{\sin \angle CDE}.$$

However, DA is the D-symmedian of triangle DEF, since the tangents at E and F with respect to the incircle meet at A. Hence, we conclude that

$$\frac{\sin \angle FDA}{\sin \angle EDA} = \frac{FD}{ED} = \frac{\sin \angle DEF}{\sin \angle DFE} = \frac{\sin \angle BDF}{\sin \angle CDE}.$$

Therefore, N is the midpoint of segment CX, as claimed. ■

Problem 86. (AOPS Forum) Let P be a point inside triangle ABC such that

$$\angle PAB + \angle PBC + \angle PCA = 90^\circ.$$

If P' is the isogonal conjugate of P, prove that PP' passes through the circumcenter of triangle ABC.

Solution. For this nice and difficult problem we first need some important facts about isogonal conjugates. Two of them are Lemma 1 and Lemma 2 from the proof of Problem 18. The third one is given by

Lemma 3. Let P be a point in the plane of the triangle ABC and let Q be its isogonal conjugate with respect to ABC. Let DEF denotes the pedal triangle of P, while XYZ the pedal triangle of Q. Then, the points D, E, F, X, Y, Z are all lying on a same circle, which is centered at the midpoint of the segment PQ.

Proof of Lemma 3. The proof uses Lemma 2 in a very clever way. More precisely, let D', E', F', X', Y', Z' be the reflections of P and Q across the sidelines of the triangle (D' and E' are the reflections of P and Q across BC, respectively, and so on). Let U be the midpoint of PQ.

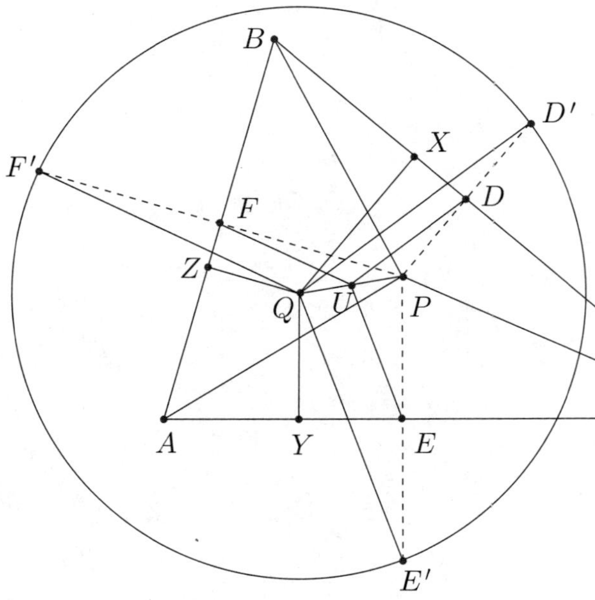

Now, note that the lines UD, UE, UF are the $U-$midlines in triangles PQD', PQE', PQF'; thus we have that

$$UD = \frac{1}{2}QD', \quad UE = \frac{1}{2}QE' \text{ and } UF = \frac{1}{2}QF'.$$

However, Theorem B tells us that Q is the circumcenter of triangle $D'E'F'$, so $QD' = QE' = QF'$, which implies that $UD = UE = UF$.

Similarly, we can deduce that $UX = UY = UF$ by looking at triangle $X'Y'Z'$ which has circumcenter P. We still need to see that for example $UD = UX$. But this is actually immediate. If U_a denotes the projection of U on BC, then the line UU_a needs to be the midline in the trapezoid $PDXQ$ since $PD \| UU_a \| QX$ and U is the midpoint of PQ; hence triangle UDX is isosceles and so $UD = UX$, as claimed. Combining this with what we found above, it follows that $UD = UE = UF = UX = UY = UZ$, so we can conclude that the points D, E, F, X, Y, Z are all lying on a same circle, which is centered at the midpoint of the segment PQ. This completes the proof.

□

Let us now return to our problem and make the following observation.

Claim. Let X, Y, Z be the projections of P on the sides BC, CA, AB of triangle ABC and let A', B', C' be the intersections of rays AP, BP, CP with the circumcircle of ABC. Then, triangles XYZ And $A'B'C'$ are homothetic.

Proof. Let $B'C'$ intersect AC at U. Using directed angles, we can write

$$\angle AUC' = \angle CAB' + \angle AB'C' = \angle PBC + \angle PCA$$

and

$$\angle AYZ = 90^\circ - \angle PYZ = 90^\circ - \angle PAB.$$

Since $\angle PAB + \angle PBC + \angle PCA = 90^\circ$, it follows that $\angle AYZ = \angle AUC'$, which gives $B'C' \| YZ$. Analogously, we get that $C'A' \| ZX$ and $A'B' \| XY$. This proves our claim.

From the Claim, we know that the pedal triangle XYZ and circumcevian triangle $A'B'C'$ of P are homothetic through a point H. Also, by Lemma 1 (of Problem 18), the perpendiculars from A, B, C to $XY \parallel B'C'$, $ZX \parallel C'A'$ and $XY \parallel A'B'$ concur at the isogonal conjugate P' of P. This means that triangles ABC and $A'B'C'$ are orthologic, with orthology centers at P' and Q, and perspective through P; therefore, by Sondat's theorem, it follows that Q, P and P' are collinear. Furthermore, since $QA' \parallel PX$, $QB' \parallel PY$, $QC' \parallel PZ$, then Q, P and H are collinear. However, by Lemma 2 (of Problem 18) the isogonal conjugate P' of P with respect to ABC is the reflection of P across the circumcenter of XYZ; hence, the line HPP' passes through the circumcenter O of triangle $A'B'C'$. This completes the proof.

■

Problem 87. (Cosmin Pohoata, Romania IMO TST 2014) Let the tangents to the circumcircle of a triangle ABC at the vertices B and C intersect each other at a point X. Consider the circle $\mathcal{X}$ centered at X with radius XB, and let M be the point of intersection of the internal angle bisector of angle A with $\mathcal{X}$ such that M lies in the interior of triangle ABC. If O is the circumcenter of ABC, denote by P the intersection of OM with the sideline BC, and let E, F be the orthogonal projections of M on CA, AB, respectively. Prove that the lines PE and FP are perpendicular.

Solution. Consider S the point of intersection of the sideline AB with circle $\mathcal{X}$. We have that $\angle BSC = 90^\circ - \angle A$, which means that the line CS is perpendicular to the line CA, and thus parallel to ME. This yields $\angle MBF = \angle MCS = \angle CME$; therefore, triangles MBF and CME are similar. This yields

$$\frac{MB}{MC} = \frac{MF}{CE} = \frac{BF}{ME},$$

and so,

$$\left(\frac{MB}{MC}\right)^2 = \frac{MF}{CE} \cdot \frac{BF}{ME} = \frac{BF}{CE}.$$

On other hand, the line OM is the M-symmedian of triangle MBC; consequently, it follows that

$$\frac{PB}{PC} = \frac{MB^2}{MC^2}.$$

Therefore,

$$\frac{PB}{PC} \cdot \frac{EC}{FB} = \frac{PB}{PC} \cdot \frac{EC}{EA} \cdot \frac{FA}{FB} = 1,$$

which, by Ceva's theorem, means that the lines AP, BE, CF are concurrent.

Next, denote by K, Q, N the intersections of the line AM with PE, FP, EF, respectively. Since the lines AP, BE, CF are concurrent, the pencil $K(E,\ A,\ F,\ B)$ is harmonic, and because the line KA passes through the midpoint of segment EF, the lines BK and EF are parallel. Thus, the quadrilateral $BKMF$ is cyclic, and so, $\angle MBD = \angle NKF = \angle NKE$. Similarly, the quadrilateral $CEMQ$ is cyclic, and thus, $\angle MCE = \angle MQE = \angle MQF$. But

$$\angle NKE + \angle NEK = 90^\circ, \quad \angle MBF + \angle MCE = 90^\circ,$$

we have that $\angle NEK = \angle PQK$. Hence, the points E, N, P, Q are concyclic, which means that the angles $\angle QPE$ and $\angle QNE$ are right angles. This completes the proof.

■

Problem 88. (Cosmin Pohoata, USAMO 2014) Let ABC be a triangle with orthocenter H and let P be the second intersection of the circumcircle of triangle AHC with the internal bisector of the angle $\angle BAC$. Let X be the circumcenter of triangle APB and Y the orthocenter of triangle APC. Prove that the length of segment XY is equal to the circumradius of triangle ABC.

Solution. It is well known that the reflection H' of the orthocenter H in the line AC lies on the circumcircle of triangle ABC. Hence, the circumcenter of triangle CAH' coincides with the circumcenter of triangle ABC. But since H' is the reflection of H in the line AC, the triangles ACH and CAH' are symmetric with respect to BC, and the circumcenter O' of triangle ACH must be the reflection of the circumcenter of triangle CAH' in the line BC, i.e. the reflection of the circumcenter of triangle ABC in the line CA.

Now since the quadrilateral $AHPC$ is cyclic and since H, Y are the orthocenters of triangles ABC, and APC, respectively, we have that

$$\angle ABC = 180^\circ - \angle AHC = 180^\circ - \angle APC = \angle AYC.$$

Hence the point Y lies on the circumcircle of triangle ABC, and therefore $OC = OY = R$, where R denotes the circumradius of triangle ABC.

On the other hand, note that the lines OX, XO', $O'O$ are the perpendicular bisectors of the segments AB, AP, and AC, respectively, we get

$$\angle OXO' = \angle BAP = \angle PAC = \angle XO'O.$$

Thus, $OO' = OX$. Combining this with $OC = OY$ and with the parallelism of the lines XO' and YC (note that these two lines are both perpendicular to AP), we conclude that the trapezoid $XYCO'$ is isosceles, and therefore $XY = O'C = OC = R$. This completes our proof.

■

Problem 89. (Iran IMO TST 2007) Let C_1, C_2 and C_3 be circles that do not intersect and so that none of them is inside another. Let (L_1, L_2), (L_3, L_4) and (L_5, L_6) be internal common tangents of the pairs of circles (C_1, C_2), (C_1, C_3), (C_2, C_3), respectively. Furthermore, let L_1, L_2, L_3, L_4, L_5, L_6 bound a hexagon $AC'BA'CB'$, where the vertices are labeled in cyclic order. Prove that the lines AA', BB', CC' are concurrent.

Solution. Define the intersections $X = L_1 \cap L_6$, $Y = L_2 \cap L_3$ and $Z = L_4 \cap L_5$. By applying the Ratio Lemma repeatedly,

$$\begin{aligned}
\frac{\sin A'AB}{\sin A'AC} &= \frac{AC}{AB} \cdot \frac{[AA'B]}{[AA'C]} \\
&= \frac{AC}{AB} \cdot \frac{A'B}{A'C} \cdot \frac{\sin AA'B}{\sin AA'C} \\
&= \frac{AC}{AB} \cdot \frac{A'B}{A'C} \cdot \frac{\sin AA'B}{\sin Z} \cdot \frac{\sin Z}{\sin Y} \cdot \frac{\sin Y}{\sin AA'C} \\
&= \frac{AC}{AB} \cdot \frac{A'B}{A'C} \cdot \frac{AZ}{AA'} \cdot \frac{\sin Z}{\sin Y} \cdot \frac{AA'}{AY} \\
&= \frac{AC}{AB} \cdot \frac{A'B}{A'C} \cdot \frac{AZ}{AY} \cdot \frac{\sin Z}{\sin Y}
\end{aligned}$$

Therefore,

$$\begin{aligned}
\prod_{\text{cyclic}} \frac{\sin A'AB}{\sin A'AC} &= \prod_{\text{cyclic}} \frac{A'B}{A'C} \cdot \frac{AZ}{AY} = \prod_{\text{cyclic}} \frac{A'B}{A'C} \cdot \frac{XB}{XC} \\
&= \prod_{\text{cyclic}} \frac{[O_1A'B] \cdot [O_1XB]}{[O_1A'C] \cdot [O_1XC]} \\
&= \prod_{\text{cyclic}} \left(\frac{O_1B}{O_1C}\right)^2 \cdot \prod_{\text{cyclic}} \left[\frac{\sin A'O_1B}{\sin A'O_1C} \cdot \frac{\sin XO_1B}{\sin XO_1C}\right]
\end{aligned}$$

On the other hand, we can easily prove that lines O_1A' and O_1X are isogonals with respect to angle $O_3O_1O_2$. Hence, we conclude that

$$\prod_{\text{cyclic}} \frac{\sin A'AB}{\sin A'AC} = 1,$$

and so the lines AA', BB', CC' are concurrent, by Trigonometric Ceva. This completes the proof.

■

Problem 90. (Cosmin Pohoata, Romania IMO TST 2007) Let ABC be a triangle, let E, F be the tangency points of the incircle $\Gamma(I)$ to the sides AC, respectively AB, and let M be the midpoint of the side BC. Let $N = AM \cap EF$, let γ be the circle of diameter BC, and let X, Y be the other (than B, C) intersection points of BI, respectively CI, with γ. Prove that

$$\frac{NX}{NY} = \frac{AC}{AB}.$$

Solution. We will assume $AB \leq AC$, so the solution matches a possible drawing. Let $T = EF \cap BC$ (for $AB = AC$, $T = \infty$), and D the tangency point of Γ to BC.

First, note that if $X' = BI \cap EF$ and $Y' = CI \cap EF$, one has $BX' \perp CX'$ and $CY' \perp BY'$. Indeed, this is nothing but the Claim from Problem 32, which we announced that it will pop up at least one more time. The second thing to notice is a beautiful consequence of the poles and polars theory.

Claim. In the configuration described above, one has $N = DI \cap EF$.

Proof. It is enough to prove that $NI \perp BC$. Let ℓ be the line through A, parallel to BC. Since the pencil $(AB, AM, AC, A\infty)$ is harmonic, it follows that the division (F, N, E, Z) is harmonic, where $Z = \ell \cap EF$. Therefore N lies on the polar of Z relative to circle Γ, and as $N \in EF$ (the polar of A), it follows that AZ is the polar of N relative to circle Γ; hence $NI \perp \ell$, so $NI \perp BC$. In conclusion, since $DI \perp BC$, one has $N \in DI$.

Returning to the problem, notice that the Claim implies $X \equiv X'$ and $Y \equiv Y'$; therefore $X, Y \in EF$. Since the division (T, B, D, C) is harmonic, it follows that D lies on the polar τ of T relative to circle γ. But $TM \perp p$, so $BC \perp \tau$, and since $DI \perp BC$, it follows that τ is, in fact, DI. By the proof of Lema 2, it thus follows that D, I, N are collinear. Since DN is the polar, it means the division (T, Y, N, X) is harmonic, thus the pencil

(DT, DY, DN, DX) is harmonic. But $DT \perp DN$, so DN is the angle bisector of $\angle XDY$, hence

$$\frac{NX}{NY} = \frac{DX}{DY} = \frac{\sin \angle DYX}{\sin \angle DXY}.$$

As quadrilaterals $BDIY$ and $CDIX$ are cyclic (since pairs of opposing angles are right angles), it follows that

$$\frac{1}{2}\angle ABC = \angle DBI = \angle DYI = \frac{1}{2}\angle DYX$$

(triangles CDY and CEY are congruent); so $\angle DYX = \angle ABC$. Similarly, $\angle DXY = \angle ACB$. Therefore

$$\frac{NX}{NY} = \frac{DX}{DY} = \frac{\sin \angle DYX}{\sin \angle DXY} = \frac{\sin \angle ABC}{\sin \angle ACB} = \frac{AC}{AB}.$$

This completes the proof.

■

Problem 91. (IMO 2013) Let the excircle of the triangle ABC lying opposite to A touch its side BC at the point A_1. Define the points B_1 and C_1 analogously. Suppose that the circumcenter of triangle $A_1B_1C_1$ lies on the circumcircle of triangle ABC. Prove that triangle ABC is right-angled.

Solution 1. Denote the circumcircles of the triangles ABC and $A_1B_1C_1$ by Ω and Γ, respectively. Denote the midpoint of the arc CB of Ω containing A by A_0, and define B_0 as well as C_0 analogously. By our hypothesis the center Q of Γ lies on Ω.

Lemma. One has $A_0B_1 = A_0C_1$. Moreover, the points A, A_0, B_1, and C_1 are concyclic. Finally, the points A and A_0 lie on the same side of B_1C_1. Similar statements hold for B and C.

Proof. Let us consider the case $A = A_0$ first. Then, triangle ABC is isosceles at A, which implies $AB_1 = AC_1$ while the remaining assertions of the Lemma are obvious. So let us suppose $A \neq A_0$ from now on.

By definition of A_0, we have $A_0B = A_0C$. It is also well known and easy to show that $BC_1 = CB_1$. Next, we have $\angle C_1BA_0 = \angle ABA_0 = \angle ACA_0 = \angle B_1CA_0$. Hence triangles A_0BC_1 and A_0CB_1 are congruent. This implies $A_0C_1 = A_0B_1$, establishing the first part of the Lemma. It also follows that $\angle A_0C_1A = \angle A_0B_1A$, as these are exterior angles at the corresponding vertices C_1 And B_1 of the congruent triangles A_0BC_1 and A_0CB_1. For that reason the points A, A_0, B_1 and C_1 are indeed the vertices of some cyclic quadrilateral two opposite sides of which are AA_0 and B_1C_1. This proves the Lemma.

Now we return to the solution. Evidently the points A_1, B_1, C_1 lies interior to some semicircle arc of Γ, so the triangle $A_1B_1C_1$ is obtuse-angled. Without loss of generality, we will assume that its angle at B_1 is obtuse. Thus Q and B_1 lie on different sides of A_1C_1; obviously, the same holds for the points B and B_1. So, the points Q and B are on the same side of A_1C_1.

Notice that the perpendicular bisector of A_1C_1 intersects Ω at two points lying on different sides of A_1C_1. By the first statement from the Lemma, both points B_0 and Q are among these points of intersection; since they share the same side of A_1C_1, they coincide.

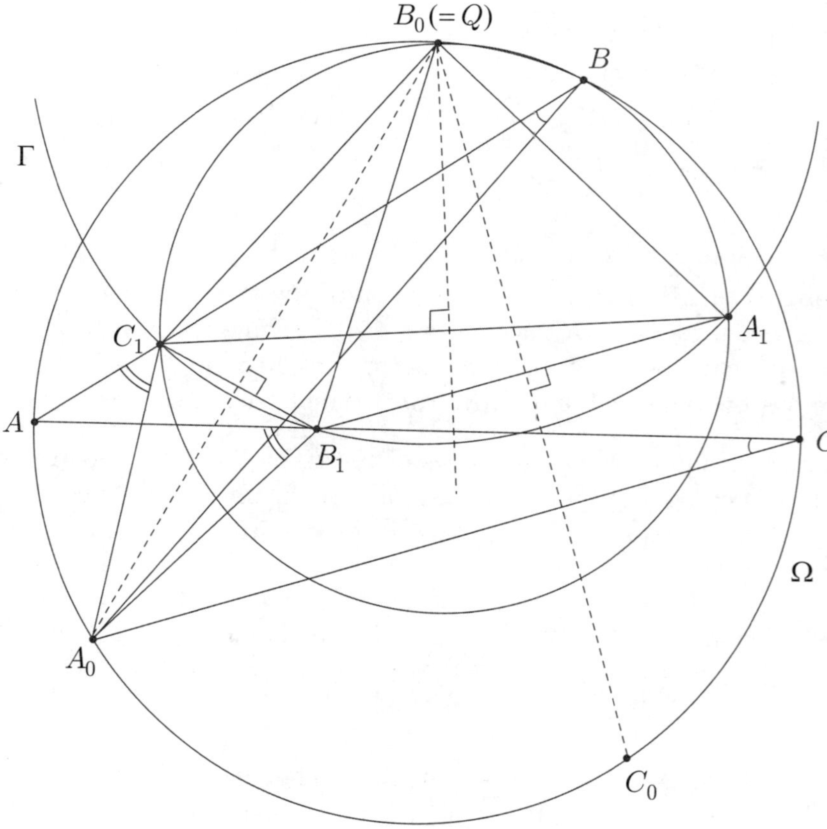

Now, by the first part of the Lemma again, the lines QA_0 and QC_0 are the perpendicular bisectors of B_1C_1 and A_1B_1, respectively. Thus

$$\begin{aligned}\angle C_1B_0A_1 &= \angle C_1B_0B_1 + \angle B_1B_0A_1 = 2\angle A_0B_0B_1 + 2\angle B_1B_0C_0 \\ &= 2\angle A_0B_0C_0 = 180^\circ - \angle ABC,\end{aligned}$$

recalling that A_0 and C_0 are the midpoints of the arcs CB and BA, respectively.

On the other hand, by the second part of the Lemma. we have

$$\angle C_1B_0A_1 = \angle C_1BA_1 = \angle ABC.$$

From the last two equalities, we get $\angle ABC = 90^\circ$, whereby the problem is solved.

Solution 2. Let Q again denote the center of the circumcircle of triangle $A_1B_1C_1$, that lies on the circumcircle Ω of triangle ABC. We first consider the case where Q coincides with one of the vertices of ABC, say $Q = B$. Then $BC_1 = BA_1$ and consequently the triangle ABC is isosceles at B. Moreover we have $BC_1 = B_1C$ in any triangle, and hence $BB_1 = BC_1 = B_1C$; similarly, $BB_1 = B_1A$. It follows that B_1 is the center of Ω and that the triangle ABC has a right angle at B.

So from now on we may suppose $Q \notin \{A, B, C\}$. We start with the following well known fact.

Lemma. Let XYZ and $X'Y'Z'$ be two triangles with $XY = X'Y'$ and $YZ = Y'Z'$.

(i) If $XZ \neq X'Z'$ and $\angle YZX = \angle Y'Z'X'$, then $\angle ZXY + \angle Z'X'Y' = 180^\circ$.

(ii) If $\angle YZX + \angle X'Z'Y' = 180^\circ$, then $\angle ZXY = \angle Y'X'Z'$.

Proof. For both parts, we may move the triangle XYZ through the plane until $Y = Y'$ and $Z = Z'$. Possibly after reflecting one of the two triangles about YZ, we may also suppose that X and X' lie on the same side of YZ if we are in the case (i) and on different sides if we are in case (ii). In both cases, the points X, Z, and X' are collinear due to the angle condition. Moreover, we have $X \neq X'$, because in case (i) we assumed $XZ \neq X'Z'$ and in case (ii) these points even lie on different sides of YZ. Thus the triangle $XX'Y$ is isosceles at Y. The claim now follows by considering the equal angles at its base.

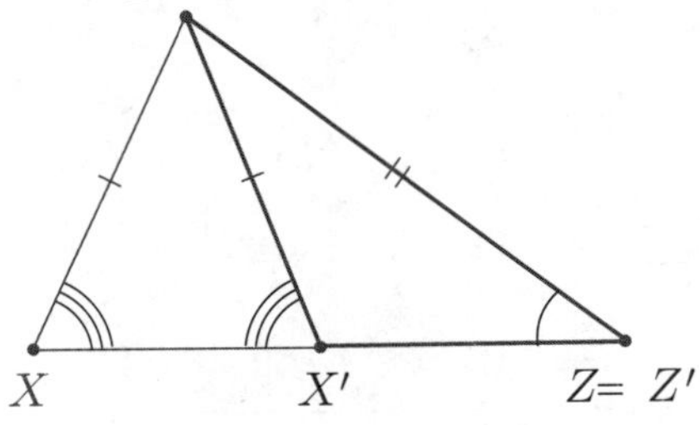

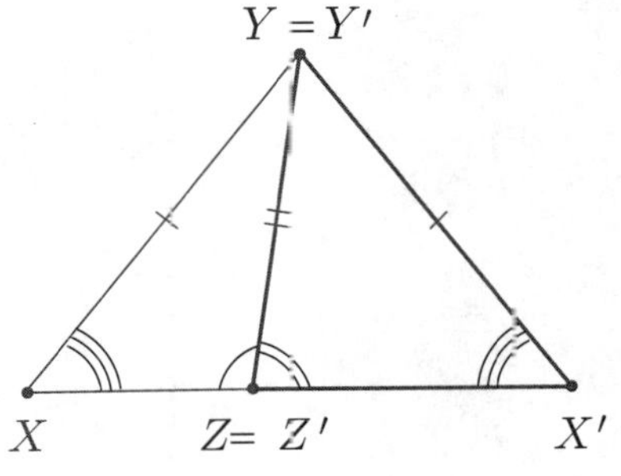

Relabeling the vertices of the triangle ABC if necessary we may suppose that Q lies in the interior of the arc AB of Ω not containing C. We will sometimes use tacitly that the six triangles QBA_1, QA_1C, QCB_1, QB_1A, QC_1A, and QBC_1 have the same orientation.

As Q cannot be the circumcenter of the triangle ABC, it is impossible that $QA = QB = QC$ and thus we may also suppose that $QC \neq QB$. Now the above Lemma (*i*) is applicable to the triangles QB_1C and QC_1B, since $QB_1 = QC_1$ and $B_1C = C_1B$, while $\angle B_1CQ = \angle C_1BQ$ holds as both angles appear over the same side of the chord QA in Ω. So we get $\angle CQB_1 + \angle BQC_1 = 180°$.

We claim that $QC = QA$. To see this, let us assume for the sake of a contradiction that $QC \neq QA$. Then arguing similarly as before but now with the triangles QA_1C and QC_1A we get

$$\angle A_1QC + \angle C_1QA = 180°.$$

Adding this equation to the one above, we get $\angle A_1QB_1 + \angle BQA = 360°$, which is absurd as both summands lie in the interval $(0°, 180°)$.

This proves $QC = QA$; so the triangles QA_1C and QC_1A are congruent, their sides being equal, which in turn yields $\angle A_1QC = \angle C_1QA$. Finally, our Lemma (*ii*) is applicable to the triangles QA_1B and QB_1A. Indeed we have $QA_1 = QB_1$ and $A_1B = B_1A$ as usual, and the angle condition $\angle A_1BQ + \angle QAB_1 = 180°$ holds as A and B lie on different sides of the chord QC in Ω. Consequently we have $\angle BQA_1 = \angle B_1QA$. This, corroborated with $\angle CQB_1 + \angle BQC_1 = 180°$, gives

$$(\angle B_1QC + \angle B_1QA) + (\angle C_1QB - \angle BQA_1) = 180°,$$

i.e. $\angle CQA + \angle A_1QC_1 = 180°$. In light of $\angle A_1QC = \angle C_1QA$, this may be rewritten as $2\angle CQA = 180°$ and as Q lies on Ω this implies that the triangle ABC has a right angle at B.

■

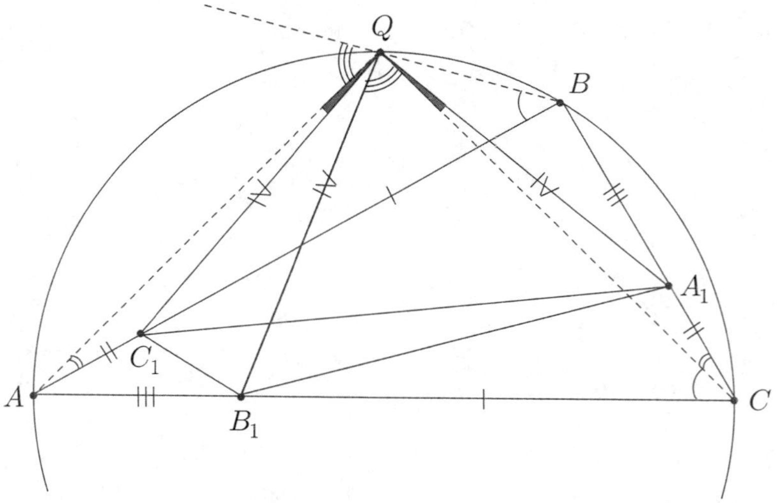

Comment. One may also check that Q is in the interior of Ω if and only if ABC is acute-angled.

Problem 92. (AOPS Forum) Let ABC be a triangle with excenters I_a, I_b, I_c and A-excircle Γ_a. Let ℓ_1, ℓ_2 be the common tangents of Γ_a and the circumcircle Ω of triangle ABC, and suppose that these meet the line BC at points D and E, respectively. Furthermore, let P and Q be the tangency points of Ω with ℓ_1 and ℓ_2, respectively. Prove that P, Q, I_b, I_c are concyclic.

Solution. We think of triangle ABC as the orthic triangle of $I_aI_bI_c$ and of Ω as its nine-point circle. Let I be the incenter of ABC and let BI and CI intersect CA, AB at points X and Y, respectively. Furthermore, let line XY intersect Ω at P_0 and Q_0, with P_0 lying on arc AB and Q_0 lying on arc AC.

The lines P_0Q_0, BC, I_bI_c concur at the foot V of the A-external bisector. Line I_bI_c is the radical axis of circles (IB_bI_c) and (BCI_bI_c), and line BC is the radical axis of circles (BCI_bI_c) and Ω. In particular, this means that V is radical center of Ω, (II_bI_c) and (BCI_bI_c); therefore P_0, Q_0, I, I_b, I_c are concyclic. We are thus left to show that $P \equiv P_0$ and $Q \equiv Q_0$.

The homothety $\mathcal{H}$ with center I_a and ratio 2 takes the nine-point circle Ω into the circumcircle $(I_aI_bI_c)$; thus Ω cuts the segment I_aQ_0 at its midpoint M. Power of Point then gives

$$I_aQ_0^2 = 2 \cdot I_aM \cdot I_aQ_0 = 4R \cdot r_a,$$

relation which we denote by $(*)$.

Now, if we let $\delta(A)$, $\delta(B)$, $\delta(C)$ be the (unsigned) distances from vertices A, B, C to the common external tangent ℓ_2 of Ω and Γ, we can apply Hartcourt's Ex-Theorem (see the Comment following the proof of Problem 29) for the excircle Γ_a and the tangent ℓ_2 (see the Comment from page):

$$2[ABC] = CA \cdot \delta(B) + AB \cdot \delta(C) - BC \cdot \delta(A).$$

However,

$$\delta(A) = \frac{QA^2}{2R}, \quad \delta(B) = \frac{QB^2}{2R}, \quad \delta(C) = \frac{QC^2}{2R};$$

thus the identity from Hartcourt becomes

$$BC \cdot CA \cdot AB = CA \cdot QB^2 + AB \cdot QC^2 - BC \cdot QA^2$$

On the other hand, however, Lagrange's theorem for points Q and I_a gives

$$CA \cdot QB^2 + AB \cdot QC^2 - BC \cdot QA^2 = (CA + AB - BC)I_aQ^2 - BC \cdot CA \cdot AB.$$

Therefore,

$$I_aQ^2 = \frac{2 \cdot BC \cdot CA \cdot AB}{CA + AB - BC} = 4R \cdot r_a.$$

Looking back at $(*)$, this proves that $I_aQ = I_aQ_0$, i.e. $Q \equiv Q_0$. Analogously, we also get that $P \equiv P_0$. This completes the proof.

■

Problem 93. (AOPS Forum) Let ABC be a triangle with circumcircle Γ and excircles Γ_a, Γ_b, Γ_c. Let T_A be the intersection of the tangents at B and C with respect to Γ. Define the points T_B and T_C similarly and let D be an arbitrary point on the side T_BT_C. Furthermore, let the tangent from D to Γ_c, which is different from the line T_BT_C, intersect the line T_CT_A at E, and let the tangent from D to Γ_b intersect T_AT_B at F. Prove that the line EF is tangent to Γ_a.

Solution. Let I_a, I_b, I_c be the excenters of triangle ABC, L be the intersection of DE with AB and N be the intersection of DF with CA. Furthermore, let ℓ be the reflection of DE with respect to I_aI_c, which meets BC and CA at L' and E', respectively. Similarly, let ℓ' be the reflection of DF with respect to I_aI_b, and suppose that this meets BC and AB at N' and F', respectively. Lastly, let I_cE' meet BC at D'.

Because

$$\begin{aligned}
\angle BD'I_c &= \angle E'L'D' + \frac{1}{2}\angle AE'L' \\
&= \frac{1}{2}(\angle E'L'D' + \angle AE'L' + \angle DLA) \\
&= \frac{1}{2}(\angle ACB + \angle DLA) \\
&= \frac{1}{2}\angle T_CDL,
\end{aligned}$$

the points D and D' are symmetric with respect to line I_bI_c. Similarly, I_bF' also passes through D', which means lines I_cE', BC and I_bF' are concurrent at D'. By Pappus's theorem, we thus get that E', I_a and F' are collinear.

On the other hand, however, points E and E' are symmetric with respect to line I_aI_c and points F and F' are symmetric with respect to I_aI_b. Hence

$$\begin{aligned}
\angle T_CEI_a + \angle T_BFI_a &= \angle AE'I_c + \angle AF'I_a \\
&= 180^\circ - \angle CAB.
\end{aligned}$$

This proves that EF is tangent to Γ_a, as claimed.

■

Problem 94. (IMO Shortlist 1999) The point M is inside the convex quadrilateral $ABCD$, such that $MA = MC$, $\angle AMB = \angle MAD + \angle MCD$, and $\angle CMD = \angle MCB + \angle MAB$. Prove that $AB \cdot CM = BC \cdot MD$ and $BM \cdot AD = MA \cdot CD$.

Solution 1. Let us construct a convex quadrilateral $PQRS$ and an interior point T such that $\triangle PTQ \sim \triangle AMB$, $\triangle QTR \sim \triangle AMD$, and $\triangle PTS \sim \triangle CMD$. We then have

$$TS = \frac{MD \cdot PT}{MC} = MD$$

and

$$\frac{TR}{TS} = \frac{TR \cdot TQ \cdot TP}{TQ \cdot TP \cdot TS} = \frac{MD \cdot MB \cdot MC}{MA \cdot MA \cdot MD} = \frac{MB}{MC},$$

where in the last equality we used the fact that $MA = MC$. We also have $\angle STR = \angle BMC$ and therefore $\triangle RTS \sim \triangle BMC$. Now the relations between the angles become

$$\angle TPS + \angle TQR = \angle PTQ \quad \text{and} \quad \angle TPQ + \angle TSR = \angle PTS,$$

implying that $PQ \| RS$ and $QR \| PS$. Hence, quadrilateral $PQRS$ is a parallelogram and hence $AB = PQ = RS$ and $QR = PS$. It follows that

$$\frac{BC}{MC} = \frac{RS}{TS} = \frac{AB}{MD},$$

which gives $AB \cdot CM = BC \cdot MD$. Furthermore,

$$\frac{AD \cdot BM}{AM} = \frac{AD \cdot QT}{AM} = QR = PS = \frac{CD \cdot TS}{MD} = CD;$$

thus, $BM \cdot AD = MA \cdot CD$. This completes the proof.

Even though the above solution is by far the best thing that you can come up with for this problem, we also include a very instructive inversion proof.

Solution 2. Invert around M with radius MA, and let us denote by X' the image of point X under this inversion. Notice that

$$\angle AD'C = \angle MD'A + \angle MD'C = \angle MAD + \angle MCD = \angle AMB$$

so we can consider point T on MB so that $\triangle AD'C \sim \triangle AMT$. This means that a spiral similarity maps $\triangle AMD' \sim \triangle ATC$. Without loss of generality, let $\angle TAM \geq \angle BAM$. These similar triangles give that $\frac{AT}{AM} = \frac{AC}{MD'}$. This, along with the ratio properties of inversion, yields that

$$\frac{AT'}{T'C} = \frac{AT}{TC} = \frac{AM}{MD'} = \frac{MD}{CM}.$$

Furthermore, the spiral similarity gives that $\frac{AM}{AT} = \frac{AD'}{AC}$, so $AC \cdot AM = AT \cdot AD'$. Moreover,

$$\frac{CD'}{MT} = \frac{AD'}{AM} \quad \text{and} \quad \frac{TC}{MD'} = \frac{AT}{AM}.$$

Thus,

$$\begin{aligned} AD' \cdot AT &= \left(\frac{CD' \cdot AM}{MT}\right)\left(\frac{AM \cdot TC}{MD'}\right) \\ &= \left(\frac{CD' \cdot MC}{MD'}\right)\left(\frac{TC \cdot CM}{MT}\right). \end{aligned}$$

The inversion however gives that $\frac{CD}{CD'} = \frac{MC}{MD'}$ and $\frac{T'C}{TC} = \frac{CM}{MT}$; thus,

$$CD \cdot T'C = AT \cdot AD' = AC \cdot CM$$

Then, $\frac{CD}{AC} = \frac{CM}{CT'} = \frac{MD}{AT'}$ so triangles CMD and $CT'A$ are similar. This means that

$$\angle BAM + \angle BCM = \angle CMD = \angle AT'C = \angle TAM + \angle TAC$$

the last one because of the inversion. We conclude that

$$\begin{aligned} \angle TAM - \angle TAB + \angle TCM - \angle TCB &= \angle BAM + \angle BCM \\ &= \angle TAC + \angle TAM. \end{aligned}$$

This means that $\angle TAB + \angle TCB = 0$. Yet, since $\angle TAM - \angle BAM \geq 0$, we see that $\angle TCM - \angle BCM \geq 0$ since T lies on BM. Then, $\angle TAB + \angle TCB \geq 0$, so the only way for equality to occur is if $T = B$. Above we had that $\frac{AT}{TC} = \frac{MD}{MC}$, so $AB \cdot MC = BC \cdot MD$. Similarly (it is not shown above, but there are mutiple ways to get this: use the symmetry of the diagram, use the ratios from the inversion and spiral similarities, or we can repeat the phantom point process for D instead of B), $BM \cdot AD = MA \cdot CD$.

■

Problem 95. (Lev Emelyanov and Tatiana Emelyanova, Forum Geometricorum) The circle passing through the feet of the internal angle bisectors of a triangle contains the Feuerbach point of the triangle.

Solution. The proof of the theorem is based on two facts: the triangle whose vertices are the feet of the internal bisectors and the the Feuerbach triangle determined by the tangency points of the incircle with the three excircles are (a) similar and (b) perspective.

Let ABC have side lengths $BC = a$, $CA = b$, $AB = c$, and circumcircle $O(R)$. Let $I_3(r_3)$ be the excircle on the side AB.

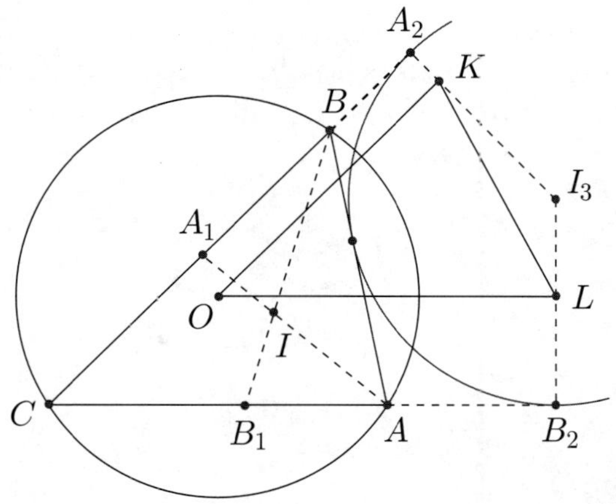

Lemma. If A_1 and B_1 are the feet of the internal angle bisectors of angles A And B, then

$$A_1B_1 = \frac{abc\sqrt{R(R+2r_3)}}{(c+a)(c+b)R}.$$

Proof of Lemma. In the diagram above, let K and L be points on I_3A_2 and I_3B_2 such that $OK \| CB$ and $OL \| CA$. Since $CA_2 = CB_2 = s$, where s is the semi-perimeter of triangle ABC,

$$OL = s - \frac{b}{2} = \frac{c+a}{2}, \qquad OK = s - \frac{a}{2} = \frac{b+c}{2}.$$

Also,

$$CB_1 = \frac{ba}{c+a}, \qquad CA_1 = \frac{ab}{b+c},$$

and

$$\frac{CB_1}{CA_1} = \frac{b+c}{c+a} = \frac{OK}{OL}.$$

Thus, triangle CA_1B_1 is similar to triangle OLK, and

$$\frac{A_1B_1}{LK} = \frac{CB_1}{OK} = \frac{2ab}{(c+a)(b+c)}.$$

Since OI_3 is a diameter of the circle through O, L, K, by the Law of Sines,

$$LK = OI_3 \sin \angle LOK = OI_3 \sin C = OI_3 \cdot \frac{c}{2R}.$$

Combining the above with Euler's formula $OI_3^2 = R(R+2r_3)$, we get the identity what we wanted.

□

We now return to the problem. Consider the nine-point circle $N(\frac{R}{2})$ tangent to the incircle at the Feuerbach point F and to excircles (I_1), (I_2), (I_3) at points F_1, F_2 and F_3, respectively (recall Feuerbach's theorem of Problem 20).

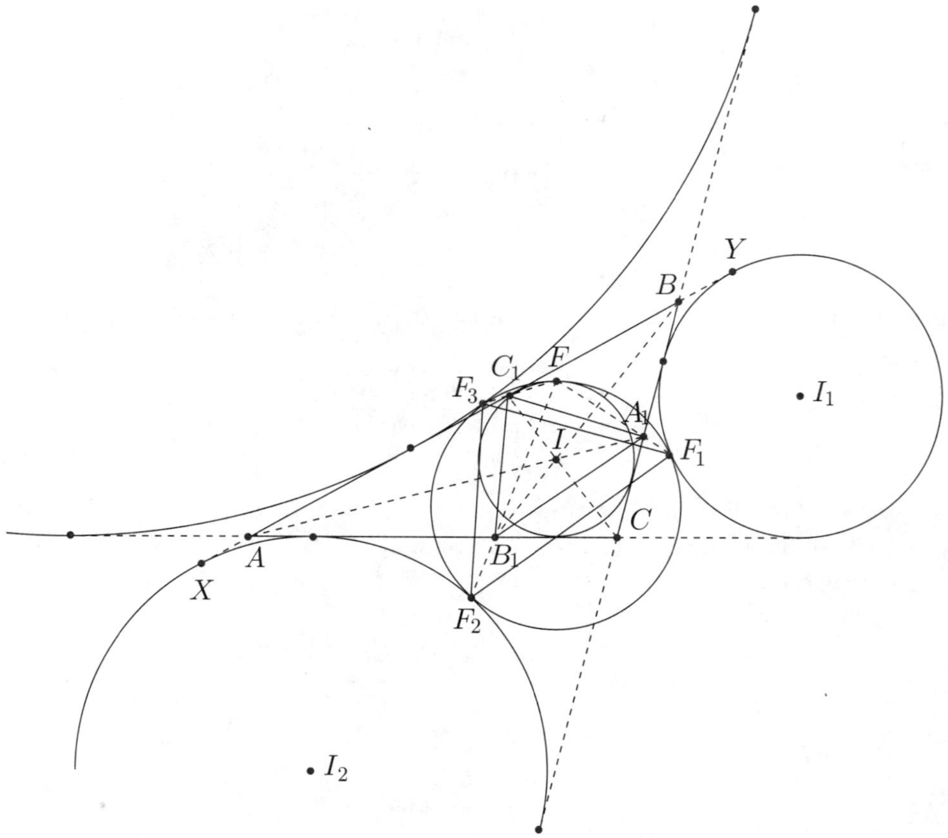

Claim 1. Triangles $A_1B_1C_1$ and $F_1F_2F_3$ are similar.

Proof. The length of the external common tangent of these two excircles is

$$XY = AY + BX - AB = s + s - c = a + b.$$

By Casey's chord theorem, we thus get

$$F_1F_2 = \frac{(a+b)\cdot\frac{R}{2}}{\sqrt{\left(\frac{R}{2}+r_1\right)\left(\frac{R}{2}+r_2\right)}} = \frac{(a+b)R}{\sqrt{(R+2r_1)(R+2r_2)}}.$$

Using the Lemma, we write

$$\frac{A_1B_1}{F_1F_2} = \frac{abc\sqrt{R(R+2r_1)(R+2r_2)(R+2r_3)}}{(a+b)(b+c)(c+a)R^2}.$$

The symmetry of this ratio in a, b, c and the exradii shows that

$$\frac{A_1B_1}{F_1F_2} = \frac{B_1C_1}{F_2F_3} = \frac{C_1A_1}{F_3F_1},$$

which shows that triangles $A_1B_1C_1$ and $F_1F_2F_3$ are similar, as claimed.

Next, note that points F, B_1, F_2 are collinear. Indeed, by the Feuerbach theorem, F is the homothetic center of the incircle and the nine-point circle, and F_2 is the internal homothetic center of the nine-point circle and the B-excircle. Moreover, B_1 is the internal homothetic center of the incircle and the B-excircle. Hence, these three homothetic centers are collinear by the Monge-D'Alembert in-theorem. Similarly, we also get that points F, C_1, F_3 are collinear and that points F, A_1, F_1 are collinear. This shows that triangles $A_1B_1C_1$ and $F_1F_2F_3$ are perspective at F, in addition to being similar. We conclude that

$$\angle C_1FA_1 + \angle C_1B_1A_1 = \angle F_3FF_1 + \angle F_3F_2F_1 = 180°,$$

which shows indeed that the circumcircle of triangle $A_1B_1C_1$ contains the Feuerbach point F. This completes the proof.

■

Comment 1. This beautiful theorem can be vastly generalized.

Generalization. Let ABC be a triangle and let P be a point in its plane. Let $A_1B_1C_1$ be its cevian triangle and let $A_2B_2C_2$ be its pedal triangle. Prove that the circumcircles of triangles $A_1B_1C_1$ and $A_2B_2C_2$ meet at a point on the nine-point circle of ABC.

The following from-the-book proof for the generalization is due to Luis Gonzalez.

Proof. The following two Lemmas proves a very important part of the problem.

Lemma 1. For any points A, B, C, D in plane, the nine-point circles of triangles ABC, BCD, CDA, DAB are concurrent.

Lemma 2. If E denotes the concurrence point from Claim 1, we furthermore have that E lies on the circumcircle of the pedal triangle of each point with respect to the triangle formed by the other three.

The common point E is known as the *Euler-Poncelet point* of the quadruplet A, B, C, D and it has amazing properties, just like the Feuerbach point.

We shall see some of them in the solution of this problem. We first prove Lemmas 1 and 2.

Proof of Lemma 1. Let the nine-point circles of triangles DAB, ABC, BCD, CDA be denoted by 1, 2, 3, and 4 like in the diagram below.

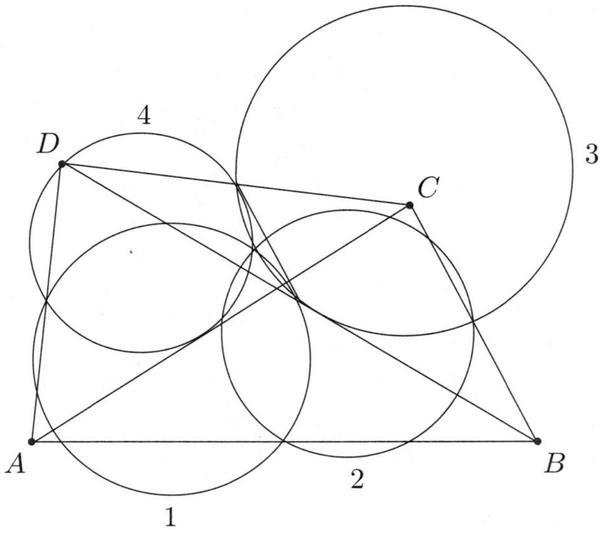

Let I, J, K, L, M, N be the midpoints of segments AB, BC, CD, DA, AC, and BD, respectively, and let E be the point of intersection of circles 3 and 4, which does not belong to segment CD.

Chasing around the midline-induced parallelisms, we write

$$\begin{aligned}\angle MEJ &= 360^\circ - \angle KEM - \angle JEK \\ &= 360^\circ - (180^\circ - \angle KLM) - \angle LMI \\ &= 180^\circ + \angle KLM - \angle LMI \\ &= 180^\circ + \angle ACD - \angle BCD \\ &= 180^\circ - \angle BCA \\ &= 180^\circ - \angle MIJ.\end{aligned}$$

Consequently, points M, E, J, I lie on the same circle. In other words, E belongs to 2 as well, and similarly, we get that E also belongs to 1. It follows that circles 1, 2, 3, 4 all pass through point E, as claimed.

□

Proof of Lemma 2. Let P, Q, R be the orthogonal projections of C on AB, BD, DA. Let M, N, L be the midpoints of segments AB, AC, AD. Nine-point circles (MNP) and (MNL) of triangles ABC and ADC intersect at N

and the Euler-Poncelet point E of $ABCD$. Thus,

$$\begin{aligned}\angle REP &= \angle REN + \angle PEN \\ &= \angle RLN + \angle PMN \\ &= \angle ADC + \angle ABC,\end{aligned}$$

whereas

$$\begin{aligned}\angle RQP &= \angle RQC + \angle PQC \\ &= \angle ADC + \angle ABC.\end{aligned}$$

Hence, $\angle REP = \angle RQP$, and so E lies on the circumcircle of triangle PQR. Analogously, we also get that E lies on each of the pedal circles of vertices A, B, D with respect to the triangle formed by the other three vertices of the quadrilateral $ABCD$, thus proving our claim.

□

Comment 2. When $ABCD$ is a cyclic quadrilateral, the Euler-Poncelet point E coincides with the center of symmetry of $ABDC$ and the quadrilateral formed by the orthocenters of triangles BCD, CDA, DAB, ABC.

We return to the proof of our generalization. Since we did not impose any convexity condition on quadrilateral $ABCD$, Lemmas 1 and 2 show that the circumcircle of triangle $A_2B_2C_2$ and the nine-point circle (N) meet at the Poncelet point P_0 of points A, B, C, and P. By the comment above, this is in fact the center of the rectangular hyperbola $\mathcal{H}$ through points A, B ,C, P, so it suffices to show that the circumcircle of triangle $A_1B_1C_1$ goes through the center of $\mathcal{H}$.

Let J be the incenter of triangle $A_1B_1C_1$ and let X, Y, Z be its excenters corresponding to vertices A_1, B_1, C_1, respectively. Regard triangle XYZ as the anticevian triangle of J with respect to triangle $A_1B_1C_1$. Consider the homology that transforms the quadrangle $ABCP$ into a square. Points A_1 and C_1 go to infinity in perpendicular directions and quadrilateral $XYZJ$ becomes a rectangle whose sides are parallel to the sides of the square $ABCP$. The image of B_1 is then the common center of $ABCP$ and $XYZJ$. By symmetry, the conic passing through A, B, C, P, and one of the vertices of the rectangle $XYZJ$, also passes through its remaining three vertices. Hence, in the original setting, points A, B, C, P, J, X, Y, Z are all on the same conic. Since J is orthocenter of triangle XYZ, it follows that this conic is the rectangular hyperbola $\mathcal{H}$, whose center is then on the nine-point circle $(A_1B_1C_1)$ of triangle XYZ. This completes the proof.

Comment 3. The result that the Euler-Poncelet point of P with respect to ABC is in fact the center of the rectangular hyperbola $\mathcal{H}$ also admits a beautiful converse.

Theorem. The cevian circumcircle of a point on a nondegenerate circumconic contains the center of the conic if and only if the conic is a rectangular hyperbola.

The proof for this fact is however slightly more involved and uses concepts from projective geometry that go beyond the scope of this book.

Problem 96. (Brown's Theorem) Let $ABCD$ be a cyclic quadrilateral. If there exists a point X such that the angles $\angle XAD$, $\angle XBA$, $\angle XCB$, $\angle XDC$ are all equal, prove that $ABCD$ is a harmonic quadrilateral.

Solution. Let us denote the sides BC, CD, DA, AB by a, b, c, d and the diagonals BD and AC by e and f, respectively. Furthermore, let Q be the area of the quadrilateral, and let R be its circumradius. Last but not least, let ω denote the common value of angles $\angle XAD$, $\angle XBA$, $\angle XCB$, $\angle XDC$.

We have that

$$\angle AXB = 180^\circ - \omega - (\angle A - \omega) = 180^\circ - \angle A.$$

Similarly, $\angle BXC = 180^\circ - \angle B$, $\angle CXD = 180^\circ - \angle B$, $\angle DXA = 180^\circ - \angle D$. Now, since

$$\frac{AX}{\sin\omega} = \frac{d}{\sin AXB} = \frac{d}{\sin(180^\circ - A)} = \frac{d}{\sin A}$$

and

$$\frac{AX}{\sin(D-\omega)} = \frac{c}{\sin AXD} = \frac{c}{\sin(180^\circ - D)} = \frac{c}{\sin D},$$

we deduce that

$$\frac{\sin(D-\omega)}{\sin\omega} = \frac{d\sin D}{c\sin A},$$

or

$$\cot\omega = \frac{d}{c\sin A} + \cot D.$$

Similarly, writing the analogues, we get

$$\cot\omega = \frac{d}{c\sin A} + \cot D = \frac{c}{b\sin D} + \cot C = \frac{b}{a\sin C} + \cot B = \frac{a}{d\sin B} + \cot A$$

or

$$\frac{d}{c\sin A} - \frac{a}{d\sin B} + \frac{b}{a\sin C} - \frac{c}{b\sin D} = \cot A - \cot B + \cot C - \cot D.$$

Since $ABCD$ is cyclic however, $\frac{1}{\sin A} = \frac{1}{\sin C}$, $\cot A = -\cot C$ etc; therefore

$$\left(\frac{d}{c}+\frac{b}{a}\right)\frac{1}{\sin A} = \left(\frac{c}{b}+\frac{a}{b}\right)\frac{1}{\sin B}.$$

But

$$(ab+cd)\sin A = (ad+bc)\sin B = 2Q;$$

hence $ac = bd$, i.e. $ABCD$ is harmonic. This completes the proof.

■

Comment. Surprisingly, the converse is also valid and can be proved on the same idea. Moreover, there is a second Brocard point X' so that $\angle X'AB = \angle X'BC = \angle X'CD = \angle X'DA = \omega'$. By simple manipulations on the above, and using Ptolemy's theorem, we obtain that

$$\cot\omega = \frac{8R^2Q}{e^2f^2}.$$

Similarly, we deduce that

$$\cot\omega' = \frac{8R^2Q}{e^2f2}.$$

Hence, $\omega = \omega'$.

Further, we will compute the distance XX'. For this, let P be the foot of the perpendicular from X to BC and $CP = x$, $PX = y$. Then

$$x = CX\cos\omega = \frac{b\sin\omega\cos\omega}{\sin C} = \frac{b(ab+cd)\cot\omega}{2Q(1+\cot^2\omega)} = \frac{bR}{c}\sin 2\omega$$

and

$$y = x\tan\omega = \frac{b(ab+cd)}{2Q(1+\cot^2\omega)} = \frac{2bR}{c}\sin^2\omega.$$

Similarly, if $x' = P'B$, $y' = P'X'$, where P' is the foot of the perpendicular from X' to BC, then

$$x' = \frac{d(ad+bc)\cot\omega}{2Q(1+\cot^2\omega}$$

and

$$y' = \frac{d(ad+bc)}{2Q(1+\cot^2\omega}.$$

Furthermore, by the Law of Cosines,

$$XX'^2 = XB^2 + X'B^2 - 2XB\cdot X'B\cdot\cos XBX'.$$

Also,

$$\sin B = \frac{ab+cd}{2bd}\cdot\tan\omega = \frac{a^2+d^2}{2ad}\cdot\tan\omega.$$

So, by using that $ac = bd$, i.e. $ef = 2ac = 2bd$ (Ptolemy's theorem), it follows that

$$XX' = \frac{efbd\cos\omega\sqrt{\cos 2\omega}}{2RQ} = \frac{b^2d^2\cos\omega\sqrt{\cos 2\omega}}{RQ}.$$

Problem 97. (Mathematics Institute Contest 2007) Let ABC be an isosceles triangle with $AB = AC$ and let M be the midpoint of BC. Find the locus of points P inside ABC for which $\angle BPM + \angle CPA = 180°$.

Solution 1. We will start with the following claim.

Claim. Prove that if, for a point P interior to the triangle, $\angle ABP = \angle BCP$, then $\angle BPM + \angle CPA = 180°$.

Notice that the configuration is fully symmetrical:

$$\angle ABP = \angle BCP \text{ if and only if } \angle ACP = \angle CBP$$

$$\angle BPM + \angle CPA = 180° \text{ if and only if } \angle CPM + \angle BPA = 180°$$

and also that $P \in AM$ guarantees the result, therefore we may assume without loss of generality (for construction's sake) that P is interior to triangle ABM. The given angle equality is readily seen to be equivalent with $P \in \Gamma$, where Γ is the arc interior to ABC of the circumcircle $\mathcal{K}$ of triangle BCI, with I the incenter of ABC. It is also immediately seen that AB and AC are tangent to $\mathcal{K}$.

Consider the Apollonius circle $\mathcal{A}$ for segment AM and ratio BA/BM; denote by U and V the points where AM meets $\mathcal{A}$. Clearly, $B, C \in \mathcal{A}$. Then

$$\frac{BA}{BM} = \frac{UA}{UM} = \frac{VA}{VM},$$

from which follows that

$$\frac{BA^2}{BM^2} = \frac{UA \cdot VA}{UM \cdot VM} = \frac{UA \cdot VA}{BM^2},$$

by symmetry and the power of a point relation, so $BA^2 = UA \cdot VA$, hence AB is tangent to $\mathcal{A}$, again by the power of a point relation. Therefore the two circles $\mathcal{K}$ and $\mathcal{A}$ coincide. Now, prolong AP until it meets $\mathcal{K}$ again at Q, and prolong PM until it meets $\mathcal{K}$ again at R. Then

$$\frac{BA}{BM} = \frac{PA}{PM} = \frac{QA}{QM},$$

therefore

$$\frac{BA^2}{BM^2} = \frac{PA \cdot QA}{PM \cdot QM} = \frac{BA^2}{PM \cdot QM},$$

by the power of a point relation, so by $BM^2 = PM \cdot QM$. But $BM^2 = PM \cdot RM$, again by the power of a point relation, so $QM = RM$, hence $BQ = CR$. It follows that $\angle BPQ = \angle CPR$, and this is enough to yield $\angle BPM + \angle CPA = 180°$.

Alternatively, one can calculate ratios using the symmetrical points P' and Q' with respect to AM, and denoting by N the meeting point of PQ' and $P'Q$. It can be obtained that $AM = AN$, thus $M = N$ and the rest easily follows as above. This settles the claim.

Returning to the problem, we will now prove that the locus is the arc Γ, together with the open segment AM. Clearly, $P \in (AM)$ fills the bill, so from now on we will assume $P \notin (AM)$. Also, as above, we will assume P is inside triangle ABM (otherwise we work with the symmetrical relations).

Assume $P \notin \Gamma$, equivalent to $\angle BPC \neq 180° - \angle ABC$, then AB and AC are not tangent to $\mathcal{K}$. Take B' and C' to be the tangency points on $\mathcal{K}$ from A, and M' to be the midpoint of $B'C'$. We are now under the conditions from the Claim (for triangle $AB'C'$), therefore $\angle B'P'M' + \angle C'PA = 180°$. But

$$\angle B'PM' = \angle BPM + \delta(\angle B'PB + \angle MPM'$$

and $\angle C'PA = \angle CPA - \delta\angle C'PC$, where $\delta = 1$ if $\angle BPC < 180° - \angle ABC$, respectively $\delta = -1$ if $\angle BPC > 180° - \angle ABC$. We have $\angle B'PB = \angle C'PC$ from the symmetry of the configuration, and $\angle BPM + \angle CPA = 180°$ given; these relations therefore imply $\angle MPM' = 0$, which can only happen when M, M' and P are collinear, i.e. $P \in AM$, which was ruled out from the start in this part of the proof. The contradiction thus reached confirms our claim.

Solution 2. Let D be the point of intersection of AP with the circumcircle of BPC and $S = DP \cap BC$. Since $\angle SPC = 180° - \angle CPA$, it follows that $\angle BPS = \angle CPM$. From Steiner's theorem applied in triangle BPC for the isogonals PS and PM, we have that

$$\frac{SB}{SC} = \frac{PB^2}{PC^2}.$$

On the other hand, using the Law of Sines, we obtain

$$\frac{SB}{SC} = \frac{DB}{DC} \cdot \frac{\sin SDB}{\sin SDC} = \frac{DB}{DC} \cdot \frac{\sin PCB}{\sin PBC} = \frac{DB}{DC} \cdot \frac{PB}{PC}.$$

Thus, by the above relation, it follows that

$$\frac{DB}{DC} = \frac{PB}{PC},$$

i.e. the quadrilateral $PBDC$ is harmonic. Therefore, the point $A' = BB \cap CC$ lies on the line PD (where by XX we denoted the tangent in X to the circumcircle of BPC).

If $A' = A$, then lines $AB = AC$ are always tangent to the circumcircle of BPC, and so the locus of P is the circle BIC, where I is the incircle of ABC. Otherwise, if $A' \neq A$, then $A' = AM \cap PS \cap BB \cap CC$, due to the fact that A' lies on PD and $A = PS \cap AM$, and by maintaining the condition that $A' \neq A$, we obtain that $PS \equiv AM$, therefore P lies on (AM).

■

Problem 98. Let ABC be a triangle and let M be the midpoint of BC and X be the midpoint of the A-altitude of ABC. Prove that the symmedian point of ABC lies on the line MX.

Solution 1. We begin with a classical result due to Lemoine.

Lemoine's Pedal Triangle Theorem. The symmedian point K of triangle ABC is the only point in the plane of ABC which is the centroid of its own pedal triangle.

Proof. We shall make use of the well-known fact that a point P lies on the A-symmedian of triangle ABC if and only if the ratio of the distances from P to the sides AB and AC equals $\frac{AB}{AC}$.

For the direct implication, let D, E, F be the projections of K on the sides BC, CA, AB and take X to be the intersection of DK with EF. We would like to show that X is the midpoint of EF, since after that we could just repeat the argument for EY and FZ and conclude that K is the centroid of DEF.

By the Ratio Lemma, we know that

$$\frac{XE}{XF} = \frac{KE}{KF} \cdot \frac{\sin XKE}{\sin XKF}.$$

However, K obviously lies on the A-symmedian, thus

$$\frac{KE}{KF} = \frac{\delta(K, AC)}{\delta(K, AB)} = \frac{AC}{AB}.$$

Furthermore, $\angle XKE = \angle C$ and $\angle XKF = \angle B$ since the quadrilaterals $KDCE$ and $KFBD$ are cyclic; thus, we conclude that

$$\frac{XE}{XF} = \frac{AC}{AB} \cdot \frac{\sin C}{\sin B} = \frac{AC}{AB} \cdot \frac{AB}{AC} = 1.$$

This proves that X is the midpoint of EF and settles the direct implication. As for the converse, things are essentially similar. Now, we know that K is a point having projections D, E, F so that

$$1 = \frac{XE}{XF} = \frac{KE}{KF} \cdot \frac{\sin XKE}{\sin XKF}.$$

The equalities $\angle XKE = \angle C$ and $\angle XKF = \angle B$ coming from the cyclic quadrilaterals $KDCE$ and $KFBD$ are independent of K being the symmedian point; thus, we immediately get that

$$\frac{KE}{KF} = \frac{AB}{AC}.$$

Hence, by the first paragraph, we conclude that K needs to lie on the A-symmedian; similarly we can do that for the vertices B and C; thus we get that K is the symmedian point of the triangle. This proves Lemoine's theorem.

□

Returning to the problem, let D, E, F be the projections of the symmedian point K on the sides BC, CA, AB, respectively. Let D' be the reflection of D in K, so K is the midpoint of segment DD'. Since DD' is parallel to the A-altitude, showing that K lies on the line MX is the same thing as showing that D' lies on the A-median AM. But this is now rather straightforward to see. First, notice that since K is the centroid of triangle DEF (by Lemoine's pedal triangle theorem), $FKED'$ is a parallelogram, as its diagonals bisect one another; thus $D'E \parallel KF$ and $D'F \parallel KE$. However, KF and KE are perpendicular to AB and AC, respectively; hence $D'E \parallel AF$ and $D'F \parallel AE$. It follows that D' is the orthocenter of triangle AEF; so AD' is perpendicular to EF. This yields that AD' needs is the A-median of triangle ABC. This completes the proof.

Solution 2. The second proof comes uses another beautiful theorem of Lemoine about the symmedian point.

Lemma. Let K be the symmedian point of triangle ABC and let x, y, z be the antiparallels drawn through K to the lines BC, CA, and AB, respectively. Then, the six points determined by x, y, z on the sides of ABC all lie on one circle.

This circle is known in literature as *Lemoine's first circle*. It has a series of beautiful properties related to important triangles centers. Here we shall just confine ourselves to its use in our proof.

Proof of Lemma. Recall the well known fact that the A-symmedian of a triangle ABC is the locus of the midpoints of the antiparallels to the line BC, which are intercepted by the sides AB and AC.

Let X_b, X_c be the intersections of x with CA, AB, respectively. Similarly, let Y_c, Y_a be the intersections of y with AB, BC, and Z_a, Z_b the intersections of z with BC, CA. Since the symmedian point K lies on all symmedians of triangle ABC, we know that $KX_b = KX_c, KY_c = KY_a, KZ_a = KZ_b$. Moreover, since y, z are antiparallels, we have that $\angle KZ_aY_a = \angle KY_aZ_a = \angle A$, thus triangle KY_aZ_a is isosceles, i.e. $KY_a = KZ_a$. Hence, $KY_a = KZ_a = KY_c = KZ_b$. Moreover, we can do the same thing for triangles KX_bZ_b, KY_cX_c to argue that they are isosceles, so we also have that $KX_b = KZ_b$ and $KY_c = KX_c$. Therefore, we conclude that

$$KZ_a = KY_a = KX_b = KZ_b = KY_c = KX_c,$$

so we get that all six points $X_b, X_c, Y_c, Y_a, Z_a, Z_b$ lie on one circle that is centered at K. This completes the proof of the Lemma.

□

We now return to our problem. The key idea is time is to note that the locus of the centers of the rectangles inscribed in triangle ABC, having one side on BC, is precisely the line MX. Indeed, the locus is first a line, since the rectangles inscribed in ABC with one side on BC are the homothetic images of the rectangles constructed outside the triangle ABC, having BC as the upper side - which all have centers lying on the perpendicular bisector of segment BC. Second, note that the midpoint of BC and the midpoint of A-altitude are the centers of the two degenerate inscribed rectangles (that have a side on BC), so they both belong to the locus; hence, the locus of the centers is indeed the line MX.

On the other hand, however, by the proof of the Lemma, we have that $KZ_b = KY_c = KZ_a = KY_a$, so $Z_aY_aZ_bY_c$ is a rectangle inscribed in ABC with Z_aY_a on BC with center K. Hence, K belongs to the locus described above. It thus follows that K lies on the line MX. This completes the second proof.

■

Problem 99. (Sherry Gong and Zuming Feng, USA IMO TST 2013) Let ABC be a triangle. Its excircles touch sides BC, CA, AB at D, E, F, respectively. Prove that the perimeter of triangle ABC is at most twice that of triangle DEF.

Solution 1. We consider the configuration shown in the diagram below. (Our proof uses directed lengths and can be easily modified for different configurations.)

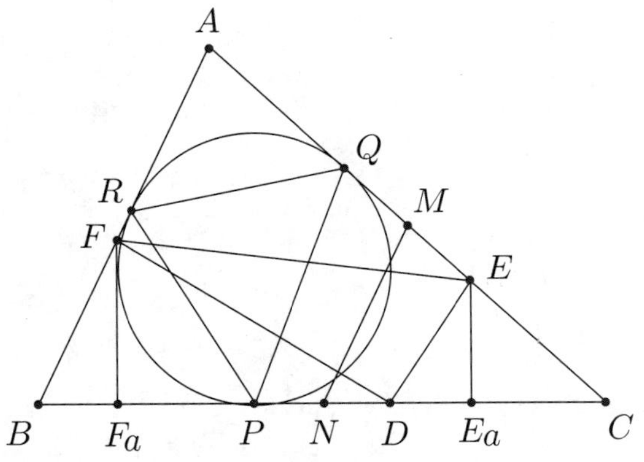

Let a, b, c denote the sides lengths of BC, CA, AB, and let A, B, C denote $\angle A$, $\angle B$, $\angle C$, respectively. Suppose that the incircle touches sides BC, CA, AB at P, Q, R, respectively. It is well known that

$$FB = CE = QA = AR = s - a,$$

where s is the semi-perimeter of ABC. Denote by E_a, F_a the feet of the perpendiculars from E, F to the line BC. It is clear that $EF \geq E_aF_a$. We therefore have

$$FE \geq F_aE_a = BC - (BF_a + E_aC) = a - (s-a)(\cos B + \cos C).$$

Summing the above inequality and its cyclic analogues yields

$$\begin{aligned} EF + FD + DE &\geq a + b + c - \sum_{\text{cyclic}} (s-a)(\cos B + \cos C) \\ &= a + b + c - (a\cos A + b\cos B + c\cos C). \end{aligned}$$

By th sum-to-product formulas, we have

$$\frac{1}{2}(\sin 2A + \sin 2B) = \sin(A+B)\cos(A-B) = \sin C\cos(A-B) \geq \sin C.$$

Summing this inequality and its cyclic analogues yields

$$\begin{aligned} \sin A + \sin B + \sin C &\geq \sin 2A + \sin 2B + \sin 2C \\ &= 2\sin A\cos A + 2\sin B\cos B + 2\sin C\cos C, \end{aligned}$$

which, by the Law of Sines, implies

$$a + b + c \geq 2a\cos A + 2b\cos B + 2c\cos C.$$

Combining this with what we got above, we conclude that

$$EF + FD + DE \geq \frac{a+b+c}{2},$$

as desired.

Solution 2. We maintain the notations of the first solution. The result clearly follows from the following two lemmas.

Lemma 1. The sum of the perimeters of triangles DEF and PQR is at least the perimeter of triangle ABC.

Proof. By symmetry, it suffices to show that $DE + PQ \geq AB$. Let M and N be the midpoints of segments CA and AB, respectively. We want to show that $DE + PQ \geq AB = 2MN$. As in the first solution, we have $CE = AQ$ and $CD = BP$ so that M and N are midpoints of segments PD and QE, respectively. Computing using vectors, we obtain

$$\begin{aligned}
\overrightarrow{MN} &= \overrightarrow{MC} + \overrightarrow{CN} \\
&= \frac{1}{2}\left(\overrightarrow{QC} + \overrightarrow{EC}\right) + \frac{1}{2}\left(\overrightarrow{CP} + \overrightarrow{CD}\right) \\
&= \frac{1}{2}\left(\overrightarrow{QC} + \overrightarrow{CP}\right) + \frac{1}{2}\left(\overrightarrow{EC} + \overrightarrow{CD}\right) \\
&= \frac{1}{2}\left(\overrightarrow{QP} + \overrightarrow{ED}\right).
\end{aligned}$$

By the triangle inequality, we have $MN \leq \frac{1}{2}(QP + ED)$, which implies

$$PQ + DE \geq 2MN = AB,$$

as claimed.

□

Lemma 2. The perimeter of triangle ABC is at least twice that of triangle PQR.

Proof. Note that $PQ = 2CP \sin \frac{C}{2}$, $QR = 2AQ \sin \frac{A}{2}$, and $RP = 2BR \sin \frac{B}{2}$. The perimeter of triangle PQR is thus equal to

$$\mathcal{P}_{PQR} = (b + c - a) \sin \frac{A}{2} + (c + a - b) \sin \frac{B}{2} + (a + b - c) \sin \frac{C}{2}.$$

By symmetry, we may assume that $a \leq b \leq c$. This yields the orderings

$$A \leq B \leq C, \quad b + c - a \geq c + a - b \geq a + b - c,$$

and

$$\sin \frac{A}{2} \leq \sin \frac{B}{2} \leq \sin \frac{C}{2}.$$

By Chebyshev's inequality, we thus have

$$\mathcal{P}_{PQR} \leq \frac{1}{3}\left((b+c-a)+(c+a-b)+(a+b-c)\right)\left(\sin\frac{A}{2}+\sin\frac{B}{2}+\sin\frac{C}{2}\right),$$

so it suffices to show that

$$\sin\frac{A}{2}+\sin\frac{B}{2}+\sin\frac{C}{2} \leq \frac{3}{2}.$$

But this follows, for instance, from Jensen's inequality applied for $y = \sin x$ (which is concave for $0 \leq x \leq 90°$).

□

Solution 3. Consider the following diagram, which contains several copies of ABC rotated and translated so that triangles ABC, $A_2B_2C_2$, $A_4B_4C_4$, and $A_6B_6C_6$ are congruent and B, $C = A_2$, $B_2 = C_4$, $A_4 = B_6$, and C_6 are collinear. Define D_i, E_i, and F_i to be the images of D, E, and F in A_iB_iC.

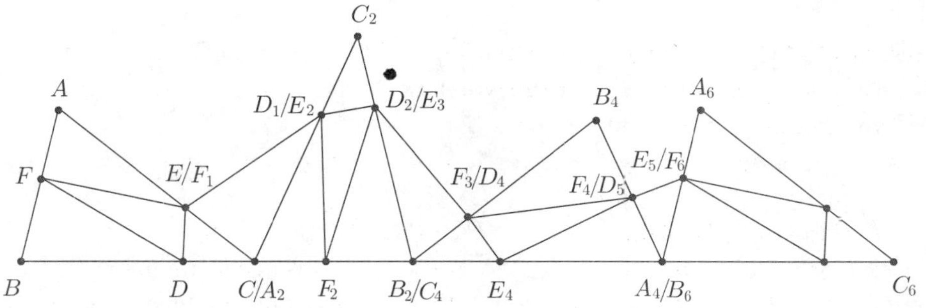

Observe that $\triangle ECE_2 \sim \triangle FBD$, $\triangle D_2B_2D_4 \sim \triangle EAF$, and $\triangle F_4A_4F_6 \sim \triangle DAE$. Using these congruences, the triangle inequality, and the fact that $FD \| F_6D_6$, we obtain

$$\begin{aligned} 2(DE+EF+FA) &\geq FE+EE_2+E_2D_2+D_2D_4+D_4F_4+F_4F_6 \\ &\geq FF_6 \\ &= AB+BC+CA. \end{aligned}$$

This completes the proof.

■

Comment. The last proof is inspired by a classical solution that Hermann Schwarz gave for

Fagnano's Theorem. For a given acute triangle, the triangle determined by the feet of its altitudes represents the inscribed triangle of minimal perimeter.

We leave the reader the pleasure of (re)discovering this proof by providing only the following picture:

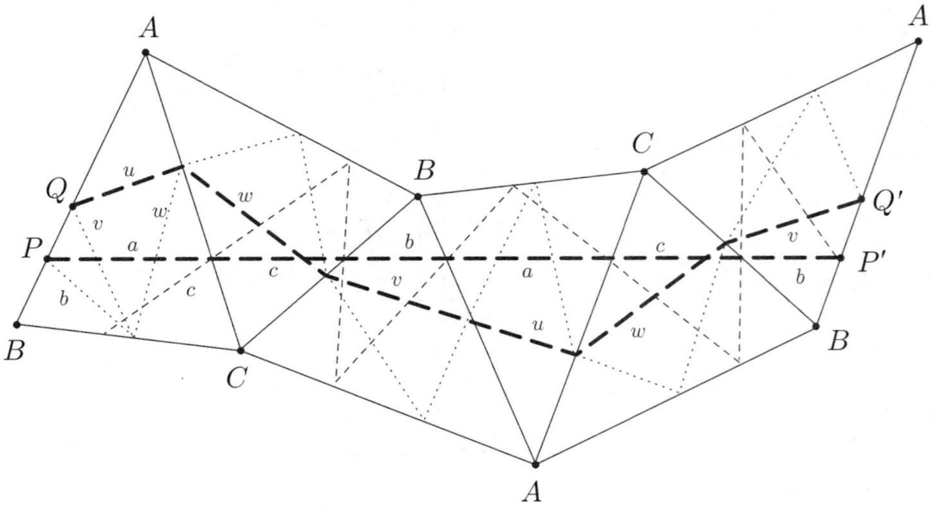

Problem 100. (Sondat's Theorem) Let ABC and $A'B'C'$ be two triangles such that the perpendiculars from vertices A, B, C to the sides $B'C'$, $C'A'$, $A'B'$ are concurrent at some point O. Prove that:

(a) Perpendicular lines from vertices A', B', C' to the sides BC, CA, AB are concurrent at some point O'.

(b) If $O = O'$, then lines AA', BB', CC' are concurrent.

(c) If $O \neq O'$, but the lines AA', BB', CC' are still concurrent at some point P, then line OO' passes through P and is perpendicular to the perspectrix of triangles ABC and $A'B'C'$.

Solution. The problem is known by geometers as *Sondat's theorem*, and is quite notorious for its difficulty. We shall present a beautiful synthetic after Jean-Louis Ayme.

(a) Since the perpendiculars from vertices A, B, C to the sides $B'C'$, $C'A'$, $A'B'$ are concurrent, Carnot's theorem yields

$$(B'A^2 - C'A^2) + (C'B^2 - A'B^2) + (A'C^2 - B'C^2) = 0.$$

However, we can rewrite this as

$$(A'B^2 - A'C^2) + (B'C^2 - B'A^2) + (C'A^2 - C'B^2) = 0,$$

which, by the same Carnot's theorem, implies that the perpendicular lines from vertices A', B', C' to the sides BC, CA, AB are concurrent.

Alternatively, one could notice that the two concurrence conditions coming from Trigonometric Ceva are equivalent.

Comment. Points O and O' are known as the *orthology centers* of triangles ABC and $A'B'C'$. The triangles themselves, if in this setting, are called orthologic triangles and they have numerous interesting properties. Parts (b) and (c) serve as examples - together they represent the so-called Sondat theorem that is quite useful in applications.

(b) We begin with a beautiful collinearity theorem due to Nikolaos Dergiades.

Lemma. Let ABC be a triangle and let Γ_a, Γ_b, Γ_c be circles having the sides BC, CA, AB as chords, respectively. Let D be the second intersection of Γ_b and Γ_c, E the second intersection of Γ_c and Γ_a, and F the second intersection of Γ_a and Γ_b. Furthermore, let the perpendicular line to AD passing through D intersect the line BC at X. Similarly, define Y and Z. Prove that X, Y, Z are collinear.

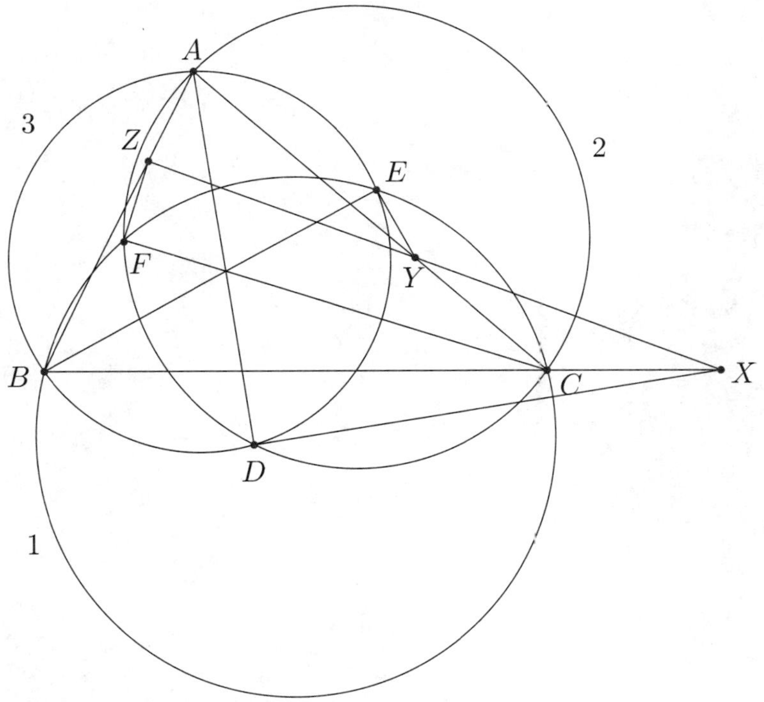

Proof. Let A', B', C' be the centers of circles Γ_a, Γ_b, Γ_c, let A'', B'', C'' be the midpoints of sides BC, CA, AB, and let a, b, c be the midpoints of segments AX, BY, CZ, respectively.

Line $B'C'$ is the perpendicular bisector of segment AD, in particular $B'C' \perp AD$. However, $AD \perp DX$, so $B'C' \| DX$, which implies that $B'C'$ passes through the midpoint a of AX. Similarly, we also get that $b \in C'A'$

and $c \in A'B'$. On the other hand, points a, b, c also lie on the midlines $B''C''$, $C''A''$, and $A''B''$, respectively, so $a = B'C' \cap B''C''$, $b = C'A' \cap C''A''$, $c = A'B' \cap A''B''$. However, the lines $A'A''$, $B'B''$, $C'C''$ are the perpendicular bisectors of sides BC, CA, AB, respectively, so in particular, they concur at the circumcenter O of triangle ABC. Desargues' theorem thus yields that a, b, and c are collinear.

We now consider quadrilateral $BCYZ$ and notice that points a, b, c are precisely the midpoints of diagonals AX, BY, and CZ, respectively. The Newton-Gauss theorem thus yields that X, Y, Z are collinear, thus proving the Lemma.

□

Let D, E, F be the intersection points of AO, BO, CO with lines $B'C'$, $C'A'$, and $A'B'$, respectively. Furthermore, let $B_{\#}$ be the intersection of BE with $A'B'$, and $C_{\#}$ the intersection of CF with $A'C'$.

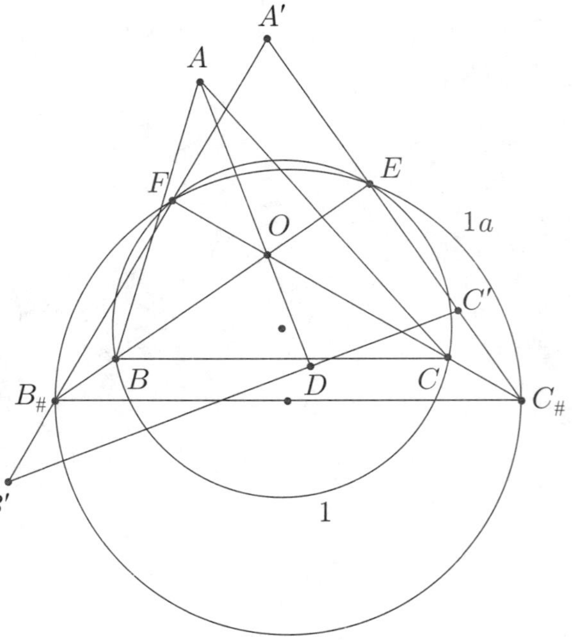

Point O is the orthocenter of triangle $A'B_{\#}C_{\#}$, so $B_{\#}$, $C_{\#}$, E, F all lie on the circle of diameter $B_{\#}C_{\#}$. In particular, $B_{\#}C_{\#}$ is an antiparallel to line EF in triangle $B_{\#}OC_{\#}$. On the other hand, $B_{\#}C_{\#} \perp A'O$ and $A'O \perp BC$ (since $O = O'$), hence $B_{\#}C_{\#} \| BC$. Thus, BC is an antiparallel to EF in triangle $B_{\#}OC_{\#}$ as well, which means that quadrilateral $BCEF$ is cyclic. Similarly, we get that quadrilaterals $CAFD$ and $ABDE$ are also cyclic, and so by the Lemma we proved, the intersections $X = B'C' \cap BC$, $Y = C'A' \cap CA$,

$Z = A'B' \cap AB$ are collinear. But then Desargues' theorem yields that lines AA', BB', and CC' are concurrent, so the proof for (b) is complete.

(c) We begin with an easy homothety Lemma.

Lemma. Let ABC and $A'B'C'$ be two triangles that are perspective at P. Let $A''B''C''$ be a triangle that is homothetic to triangle $A'B'C'$, with homothety center P. Let $X = BC \cap B'C'$, $Y = CA \cap C'A'$, $Z = AB \cap A'B'$, and let $X' = BC \cap B''C''$, $Y' = CA \cap C''A''$, $Z' = AB \cap A''B''$. Then, lines XYZ and $X''Y''Z''$ are parallel.

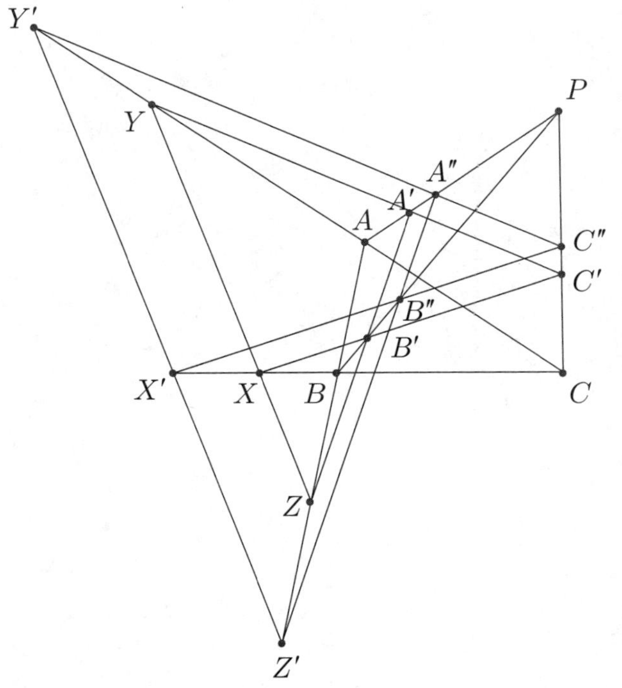

We omit the proof and leave it as an Desargues theorem practice exercise for the reader. We prepare the application of this Lemma with two preliminary claims.

Claim 1. Suppose ABC and $A'B'C'$ are orthologic triangles with $O = O'$ and so that lines AA', BB', CC' are concurrent. Let X be the intersection of BC with $B'C'$. Then, lines OX and AA' are perpendicular.

Proof. Let D be the intersection of AO with $B'C'$, E the intersection of BO with $C'A'$, and F the intersection of CO with $A'B'$. By the proof of (b), points B, C, E, F are concyclic and points A, C, F, D are concyclic. Furthermore, if D' is the intersection of $A'O$ with BC and E' is the intersection of $B'O$ with CA, points A', B', D', E' are also concyclic. Power of Point O thus yields

$$OA' \cdot OD' = OB' \cdot OE' = OC \cdot OF = OA \cdot OD.$$

Hence, $OA' \cdot OD' = OA \cdot OD$, i.e. points A, A', D, and D' are concyclic.

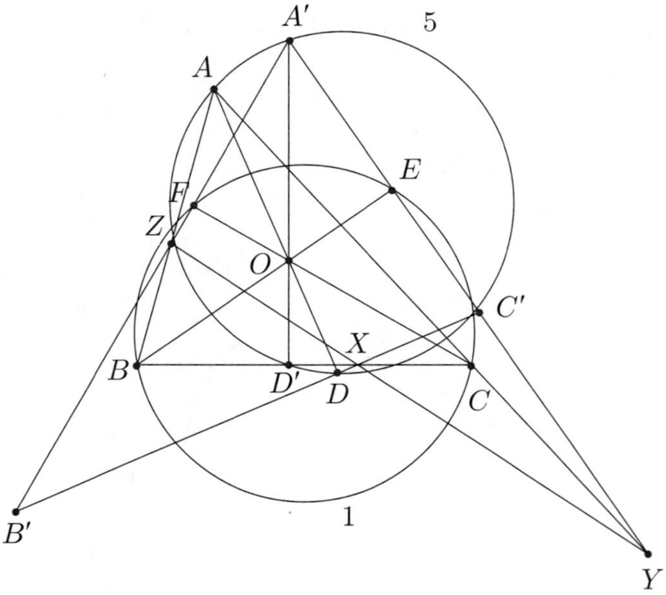

On the other hand, the circle γ of diameter OX passes through points D and D' and so it is easy to check that the tangent τ at O with respect to γ is parallel to AA'. But τ is also perpendicular to OX; hence $OX \perp AA'$, as desired.

Claim 2. Using the same notations and hypothesis from Claim 1, furthermore let P be the common point of lines AA', BB', CC', Y be the intersection of CA with $C'A'$, and Z be the intersection of AB with $A'B'$. Then, OP is perpendicular to the line passing through X, Y, and Z.

Proof. Let D' be the intersection of $A'O$ with BC and let X' be the intersection of OX with AA'. Circle of diameter $A'X$ passes through D' and X' and so $OX \cdot OX' = OA' \cdot OD'$.

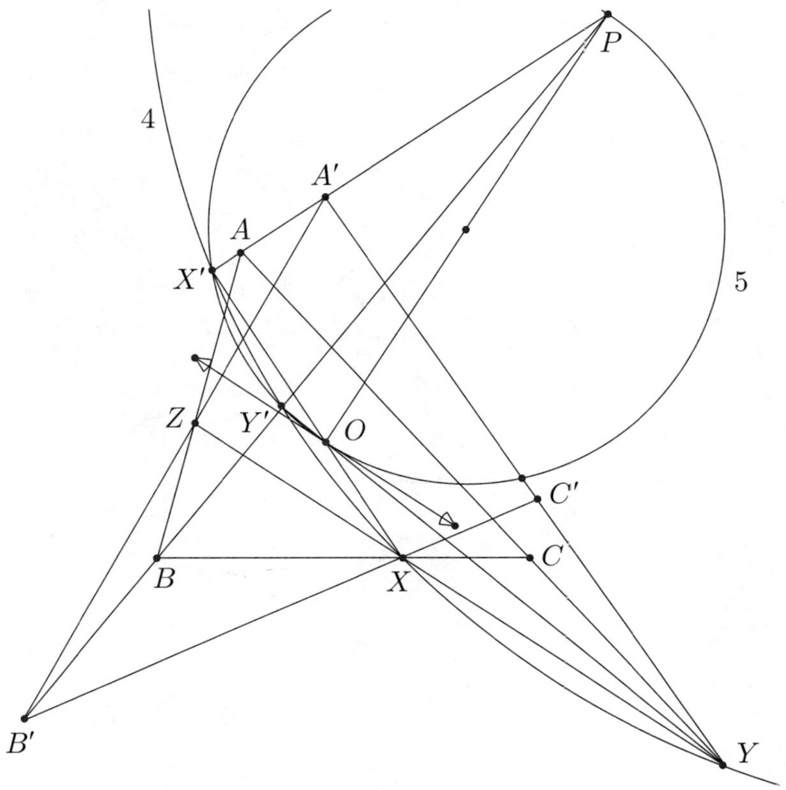

Let $B'O$ meet CA at E'. By the proof of (b), points A', B', D', E' are concyclic; therefore,

$$OA' \cdot OD' = OB' \cdot OE'.$$

But the circle of diameter $B'Y$ passes through E' and the intersection Y' of OY and BB'. Hence,

$$OB' \cdot OE' = OY' \cdot OY.$$

Combining the three Power of Point identities above, we thus get that $OX \cdot OX' = OY \cdot OY'$, so X, X', Y, Y' are concyclic.

We conclude in the same style as we conclude the proof of Claim 1. Circle ω of diameter OP passes through X' and Y', and it is easy to check that the tangent at O with respect to ω is parallel to XY. Consequently, $XY \perp OP$, which proves our claim.

We are finally ready to see what happens when $O \neq O'$. We maintain the notations from Claim 2 above. Furthermore, let A^* be the intersection of the perpendicular from O to BC with the line AA', B^* the intersection of the parallel to $A'B'$ through A^* with BB', and C^* the intersection of the parallel to $B'C'$ through B^* with CC'.

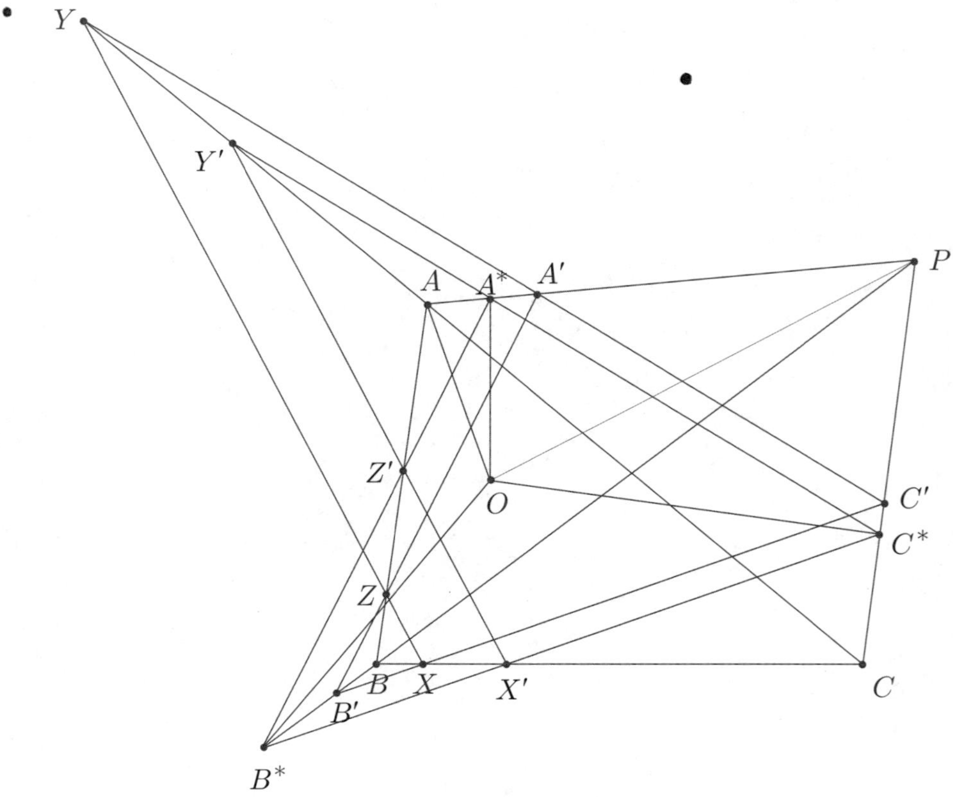

By Desargues' theorem, applied for triangles $A^*B^*C^*$ and $A'B'C'$, which are perspective at P, the lines A^*C^* and $A'C'$ are parallel; thus, triangles $A^*B^*C^*$ and $A'B'C'$ are homothetic, with homothety center P. In particular, since $AO \perp B'C' \| B^*C^*$, it follows that $AO \perp B^*C^*$, and the analogous relations. In other words, triangles ABC and $A^*B^*C^*$ are orthologic too. But magic happened! Unlike ABC and $A'B'C'$, their orthology centers coincide at O, while still being perspective at P, by construction. It thus follows from Claim 2 that the points $X' = B^*C^* \cap BC$, $Y' = C^*A^* \cap CA$, $Z' = A^*B^* \cap AB$ are collinear on a line that is perpendicular to OP.

We are practically done, for we can sense that the Lemma is around the corner. In the same way, we can define this time A^* to be the intersection of AA' with the perpendicular from O' to $B'C'$, B^* the intersection of BB' with the parallel to AB through A^*, and C^* the intersection of CC' with the parallel to BC through B^*. Similarly, we get that triangles ABC and $A^*B^*C^*$ are homothetic, with homothety center P, and consequently that triangles $A'B'C'$ and $A^*B^*C^*$ are orthologic with coincident orthology center O', while being perspective at P, by construction. By Claim 2, it follows again that $X'' = B^*C^* \cap B'C'$, $Y'' = C^*A^* \cap C'A'$, $Z'' = A^*B^* \cap A'B'$ are collinear on a line that is perpendicular to $O'P$. But the Lemma we began with tells us that $XYZ \| X''Y''Z''$. Hence, from the above, it follows that $OP \| OP'$. In particular, this means that points O, O', and P are collinear on a line that is perpendicular to line XYZ. This completes the proof of Sondat's theorem.

■

Problem 101. (Cosmin Pohoata, Forum Geometricorum) Let ABC be a non-equilateral triangle with circumcircle Γ and let A', B', C' be the reflections of A, B, C across the sides BC, CA, AB, respectively. Let A_t, B_t, C_t be the pairwise intersections of the tangents at the vertices of ABC with respect to Γ. Prove that the circumcircles of triangles $A_tB'C'$, $B_tC'A'$, $C_tA'B'$ are concurrent.

Solution. We shall make use of the notion of directed angles (ℓ_1, ℓ_2) between two lines ℓ_1 and ℓ_2 as the angle of rotation (defined modulo 180°) that will bring ℓ_1 to ℓ_2 in the same orientation as BC.

Let α, β, γ be lines through the lines A, B, C, respectively parallel the Euler line. By the result of Problem 48, their reflections α', β', γ' in the sides BC, CA, AB pass through the Parry reflection point P.

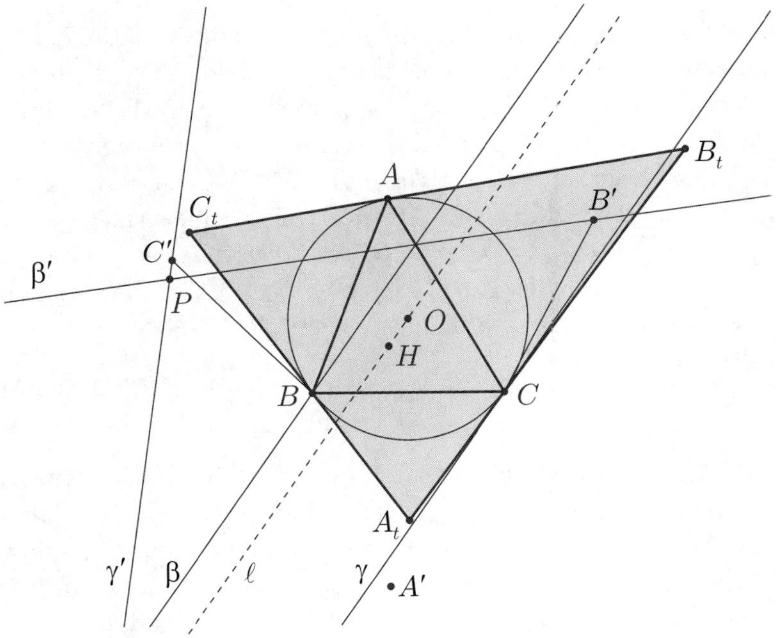

Now, since α, β, γ are parallel,

$$\begin{aligned}
(PB', PC') &= (\beta', \gamma') \\
&= (\beta', BC) + (BC, \gamma') \\
&= -(\beta', B'C) - (BC', \gamma') \\
&= -(B'C, \beta) + (\gamma, BC') \\
&= (B'C, \beta) + (\beta, BC') \\
&= (B'C, BC') \\
&= (B'C, AC) + (AC, BC') \\
&= (AC, AB) + (AC, BC') \\
&= (AC, AB) + (AB, BC) + (AC, AB) + (AB, BC') \\
&= 2(AC, AB) \\
&= (OC, OB) \\
&= (A_tC, A_tB).
\end{aligned}$$

Since $A_tB = A_tC$ and $BC' = BC = B'C$, we conclude that the triangles A_tBC' and A_tCB' are congruent. Hence, $(A_tB', A_tC') = (A_tC, A_tB)$. This gives $(PB', PC') = (A_tB, A_tC')$, and the points P, A_t, B', C' are concyclic. The circle $A_tB'C'$ contains the Parry reflection point, so do the circles $B_tC'A'$ and $C_tA'B'$.

■

Comment. Note also that the circumcircles of triangles $A'B_tC_t$, $B'C_tA_t$ and $C'A_tB_t$ are concurrent. Indeed, consider the inversion Ψ with respect to the Parry reflection point P (and arbitrary power). Since the circumcircles of triangles $A_tB'C'$, $B_tC'A'$, $C_tA'B'$ are concurrent at P, their images under Ψ are three lines bounding the triangle $A'^*B'^*C'^*$, where A'^*, B'^*, C'^* are the inversive images of A', B', C', respectively. Since the points A_t^*, B_t^*, C_t^* lie on the lines $B'^*C'^*$, $C'^*A'^*$, $A'^*B'^*$, respectively, by Miquel's theorem, the circumcircles of triangles $A_t^*B'^*C'^*$, $B_t^*C'^*A'^*$, $C_t^*A'^*B'^*$ are concurrent. But then, so are their inversive images, the circles $A_tB'C'$, $B_tC'A'$, $C_tA'B'$, therefore the claim is proven.

Problem 102. (Nairi Sedrakyan, International Zhautykov Olympiad 2014) Four segments divide a convex quadrilateral into nine quadrilaterals as in the diagram. The points of intersections of these segments lie on the diagonals of the quadrilateral. It is known that quadrilaterals 1, 2, 3, 4 admit inscribed circles. Prove that the quadrilateral 5 also has an inscribed circle.

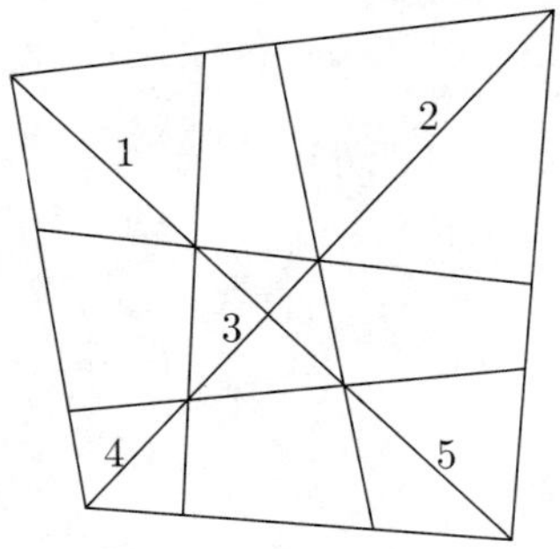

Solution. The key idea is to make use of the following result due to Marius Iosifescu repeatedly.

Iosifescu's Theorem. A convex quadrilateral $ABCD$ admits an inscribed circle if and only if

$$\tan\frac{x}{2}\cdot\tan\frac{z}{2}=\tan\frac{y}{2}\cdot\tan\frac{w}{2},$$

where $x=\angle ABD$, $y=\angle ADB$, $z=\angle BDC$ and $w=\angle DBC$.

Proof. Using the trigonometric formula

$$\tan^2\frac{u}{2}=\frac{1-\cos u}{1+\cos u},$$

we get that the equality in the theory is equivalent to

$$(1-\cos x)(1-\cos z)(1+\cos y)(1+\cos w)$$

$$=(1-\cos y)(1-\cos w)(1+\cos x)(1+\cos z).$$

Let $a=AB$, $b=BC$, $c=CD$, $d=DA$ and $q=BD$. From the Law of Cosines, we have

$$\cos x=\frac{a^2+q^2-d^2}{2aq},$$

so that

$$1-\cos x=\frac{d^2-(a-q)^2}{2aq}=\frac{(d+a-q)(d-a+q)}{2aq}$$

and

$$1+\cos x=\frac{(a+q)^2-d^2}{2aq}=\frac{(a+q+d)(a+q-d)}{2aq}.$$

In the same way

$$1-\cos y=\frac{(a+d-q)(a-d+q)}{2dq},\quad 1+\cos y=\frac{(d+q+a)(d+q-a)}{2dq},$$

$$1-\cos z=\frac{(b+c-q)(b-c+q)}{2cq}, \quad 1+\cos z=\frac{(c+q+b)(c+q-b)}{2cq},$$
$$1-\cos w=\frac{(c+b-q)(c-b+q)}{2bq}, \quad 1+\cos w=\frac{(b+q+c)(b+q-c)}{2bq}.$$

Thus, after some rearranging, the identity we started with becomes

$$P\left((d-a+q)^2(b-c+q)^2-(a-d+q)^2(c-b+q)^2\right)=0,$$

where

$$P=\frac{d+a-q)(b+c-q)(d+q+a)(b+q+c)}{16abcdq^4}$$

is a positive expression according to the triangle inequality applied in triangles ABD and BCD. Factoring the above and cancelling out terms in the parantheses yields

$$4qP(b+d-a-c)\left((d-a)(b-c)+q^2\right)=0,$$

where the expression in the second paranthesis can never be equal to zero - since, by the triangle inequality, $q>a-d$ and $q>b-c$, so q^2 is always either (strictly) less or greater than $(a-d)(b-c)$. It thus follows that

$$\tan\frac{x}{2}\cdot\tan\frac{z}{2}=\tan\frac{y}{2}\cdot\tan\frac{w}{2}$$

holds if and only if $b+d-a-c=0$, i.e. $b+d=a+c$. By Pithot's theorem, we therefore conclude that this happens if and only if $ABCD$ admits an inscribed circle.

□

In order to use Iosifescu's theorem for quadrilateral 5, we aim to prove that

$$\tan\frac{\angle FHG}{2}\cdot\tan\frac{\angle BDC}{2}=\tan\frac{\angle ADB}{2}\cdot\tan\frac{\angle EHF}{2}.$$

However, quadrilateral 3 has an incircle, so

$$\tan\frac{\angle EFH}{2}\cdot\tan\frac{\angle FHG}{2}=\tan\frac{\angle EHF}{2}\cdot\tan\frac{\angle HFG}{2}.$$

It thus suffices to show that

$$\tan\frac{\angle EFH}{2}\cdot\tan\frac{\angle ADB}{2}=\tan\frac{\angle HFG}{2}\cdot\tan\frac{\angle BDC}{2}.$$

However, by applying Iosifescu's lemma once again to quadrilateral 1, the above happens if and only if

$$\tan\frac{\angle ABD}{2}\cdot\tan\frac{\angle BDC}{2}=\tan\frac{\angle ADB}{2}\cdot\tan\frac{\angle DBC}{2},$$

which we know is true if and only if $ABCD$ has inscribed circle.

This is where the magic comes in. We show that $ABCD$ has an inscribed circle by using Iosifescu's theorem for the other diagonal AC of $ABCD$. Namely, it suffices to show that

$$\tan\frac{\angle BAC}{2}\cdot\tan\frac{\angle ACD}{2}=\tan\frac{\angle BCA}{2}\cdot\tan\frac{\angle CAD}{2}.$$

This is however easy to prove in the same style as above. Indeed, by writing out Iosifescu's lemma for quadrilaterals 2, 3 and 4, and keeping in mind the vertical angles from the vertices of quadrilateral 3, we get precisely the above statement. We leave the details for the reader.

■

Problem 103. (Paul Yiu, Crux Mathematicorum) Given a triangle ABC, let C_A be the circle tangent to the sides AB, AC and to the circumcircle internally. Define C_B and C_C analogously. Find the triangle, unique up to similarity, for which the inradius and the radii of the three circles C_A, C_B and C_C are in arithmetic progression.

Solution. We denote by r, r_A, r_B and r_C the radii of the incircle and the mixtilinear incircles C_A, C_B and C_C, respectively. It is known that

$$\begin{aligned} r_A &= r\sec^2\frac{A}{2}=r\left(1+\tan^2\frac{A}{2}\right),\\ r_B &= r\sec^2\frac{B}{2}=r\left(1+\tan^2\frac{B}{2}\right),\\ r_C &= r\sec^2\frac{C}{2}=r\left(1+\tan^2\frac{C}{2}\right). \end{aligned}$$

Indeed, we refer again to Comment 2 from the proof of Problem 36. The mixtilinear incircles C_A, C_B, C_C are degenerate cases of Thebault circles so the inradius identity applies.

Now, if we assume, without loss of generality, that $r\le r_A\le r_B\le r_C$, then the condition that these radii are in arithmetic progression implies that $r_A=r+d$, $r_B=r+2d$, $r_C=r+3d$ for some common difference d.

Using the formulas above, it follows that $d=r\tan^2\frac{A}{2}$, $2d=r\tan^2\frac{B}{2}$, $3d=r\tan^2\frac{C}{2}$.

From this it follows that

$$\tan\frac{B}{2}=\sqrt{2}\tan\frac{A}{2}\quad\text{and}\quad\tan\frac{C}{2}=\sqrt{3}\tan\frac{A}{2}.$$

It we let a, b and c be the sides opposite angles A, B and C, respectively, we know that we can write $a=x+y$, $b=x+z$, $c=y+Z$, where x, y, z are positive real numbers. We furthermore have that $\tan\frac{A}{2}=\frac{r}{z}$, $\tan\frac{B}{2}=\frac{r}{y}$ and $\tan\frac{C}{2}=\frac{r}{x}$; thus, we get that

$$x:y:z=\sqrt{2}:\sqrt{3}:\sqrt{6},$$

and hence, it follows that the radii are in the desired arithmetic progression if the three sides of the triangle are in the ratio

$$a:b:c=\sqrt{2}+\sqrt{3}:\sqrt{2}+\sqrt{6}:\sqrt{3}+\sqrt{6}.$$

Furthermore, we know that

$$\tan\frac{A}{2}\tan\frac{B}{2}+\tan\frac{B}{2}\tan\frac{C}{2}+\tan\frac{C}{2}\tan\frac{A}{2}=1,$$

thus, we get that

$$\tan^2\frac{A}{2}=(\sqrt{2}+\sqrt{3}+\sqrt{6})^{-1}.$$

This tells us that the angles in the triangle are

$$\begin{aligned} A &= 2\tan^{-1}\frac{1}{\sqrt{\sqrt{2}+\sqrt{3}+\sqrt{6}}}, \\ B &= 2\tan^{-1}\frac{\sqrt{2}}{\sqrt{\sqrt{2}+\sqrt{3}+\sqrt{6}}}, \\ C &= 2\tan^{-1}\frac{\sqrt{3}}{\sqrt{\sqrt{2}+\sqrt{3}+\sqrt{6}}}. \end{aligned}$$

■

Problem 104. (Kostas Vittas, AOPS Forum) Let $BCKL$, $CAHF$, $ABDE$ be rectangles constructed on the sides BC, CA, AB of triangle ABC. Let A' be the intersection of EK with HL, B' the intersection of HL with DF, and C' the intersection of DF with EK. Prove that the lines AA', BB', CC' are concurrent.

Solution. Let (O_1), (O_2), (O_3) be the circumcircles of the rectangles $ABDE$, $ACFH$, $BCKL$, respectively. Furthermore, let A'' be the second intersection of circles (O_1) and (O_2). Similarly, define B'' and C''.

Note that

$$\angle AA''D+\angle AA''F=\angle ABD+\angle ACF=180^\circ,$$

so A'' lies on the line DF. Similarly, we also get that B'' lies on EK and C'' lies on HL. Moreover, note that lines AA'' is the radical axis of (O_2) and

(O_3), BB'' is the radical axis of (O_3) and (O_1), and CC'' is the radical axis of (O_1) and (O_2). It therefore follows that AA'', BB'', C'' are concurrent at the radical center of the three circles, which we denote by P.

Looking now at triangles ABC and $A'B'C'$, we remark that the perpendiculars through vertices A, B, C to the sides $B'C'$, $C'A'$, $A'B'$ are concurrent, or, in other words ABC and $A'B'C'$ are orthologic triangles; hence, by Problem 100 (a), the perpendiculars from A', B', C' to BC, CA, AB, respectively, are also concurrent at some point, say P'. We claim that $P' = P$. This, combined with part (b) of the same Problem 100 shows that lines AA', BB', CC' are concurrent, as desired.

Circle (O_3) passes through vertices B'', C'' of triangle $PB''C''$ and intersects its side lines PB'', PC'', at points B and C, respectively. Line BC is therefore antiparallel to $B''C''$ with respect to angle $\angle P$ of triangle $PB''C''$. Moreover, since $\angle A'B''P = \angle A'C''P = 90°$, segment AP is the diameter of the circumcircle of triangle $PB''C''$; hence, it follows that $AP \perp BC$. In the same way, we also get that $BP \perp AC$ and $CP \perp AB$, which shows indeed that $P' = P$, as claimed. This completes the proof.

■

Problem 105. (Rabinowitz's Seven Circle Theorem) Let Γ be a circle in plane and let γ_A, γ_B, γ_C, γ_D, γ_E, and γ_F be six circle inside it, each internally tangent to Γ and tangent to each other in pairs. Let A, B, C, D, E, and F be the consecutive points of tangency of the small circles with the outer circle. Prove that the lines AD, BE, and CF are concurrent.

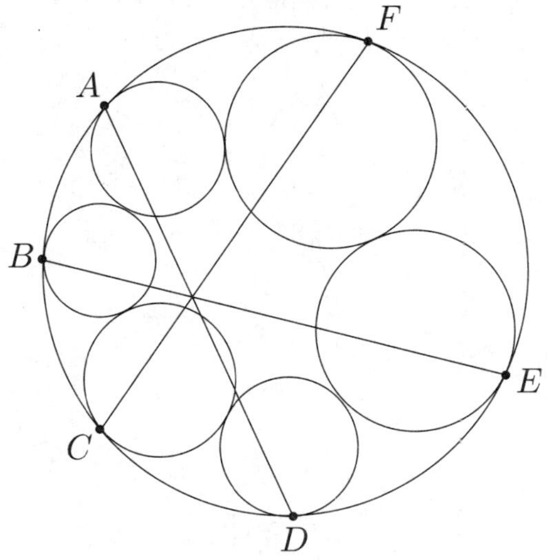

Solution. We begin with a Ceva-type theorem for hexagons.

Lemma. Let A, B, C, D, E, and F be six consecutive points around the circumference of a circle. Then, chords AD, BE, CF concur if and only if $AB \cdot CD \cdot EF = BC \cdot DE \cdot FA$.

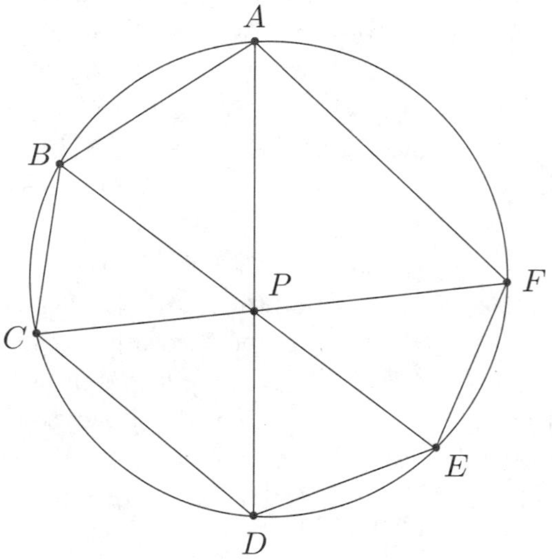

Proof. By the Law of Sines, we have that $AB = 2R\sin AEB$, $CD = 2R\sin CAD$, $EF = 2R\sin ECF$, $BC = 2R\sin BEC$, $DE = 2R\sin DAE$, $FA = 2R\sin ACF$. It follows that $AB \cdot CD \cdot EF = BC \cdot DE \cdot FA$ if and only if

$$\frac{\sin CAD}{\sin EAD} \cdot \frac{\sin AEB}{\sin CEB} \cdot \frac{\sin ECF}{\sin ACF} = 1,$$

which is equivalent to AD, BE, CF being concurrent via Trigonometric Ceva in triangle ACE. This proves the Lemma.

□

Let the radius of Γ be R and let the radius of γ_X be r_X for all $X \in \mathcal{V} = \{A, \ldots, F\}$. By a computation similar to the one we did for the Casey Chord theorem from the proof of Problem 80, we have that

$$XY = 2Rf(r_X)f(r_Y)$$

where $f(r) = \sqrt{\frac{r}{R-r}}$ and $X, Y \in \mathcal{V}$. Hence

$$AB \cdot CD \cdot EF = 8R^3 f(r_A)f(r_B)f(r_C)f(r_D)f(r_E)f(r_F) = BC \cdot DE \cdot FA.$$

Therefore, by the above Lemma, the lines AD, BE, CF are concurrent, which completes the proof.

■

Comment. As you might expect after seeing the proof, the result also has a version for the case when internal tangency is replaced (everywhere) by external tangency.

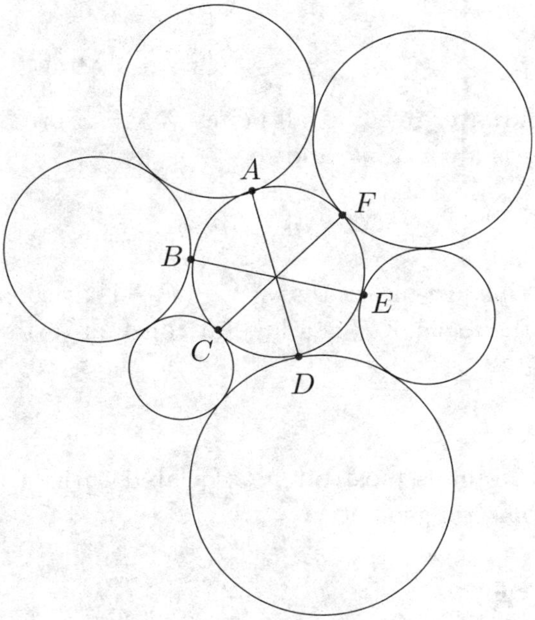

Problem 106. (Leon-Anne's Theorem) Let $ABCD$ be a convex quadrilateral and let r be a real number. Find the locus of the points P in plane for which the sum of areas $[PAB] + [PCD]$ equals r is a line.

Solution. Let P be a point satisfying $S_{PAB} + S_{PCD} = k^2$. If $ABCD$ is a rectangle, then the result is easy. We leave it as an exercise for the reader. Otherwise, let us suppose, without loss of generality, that AB meets CD at X. Now, construct the points Y and Z on the rays XA and XD so that $XY = AB$ and $XZ = CD$.

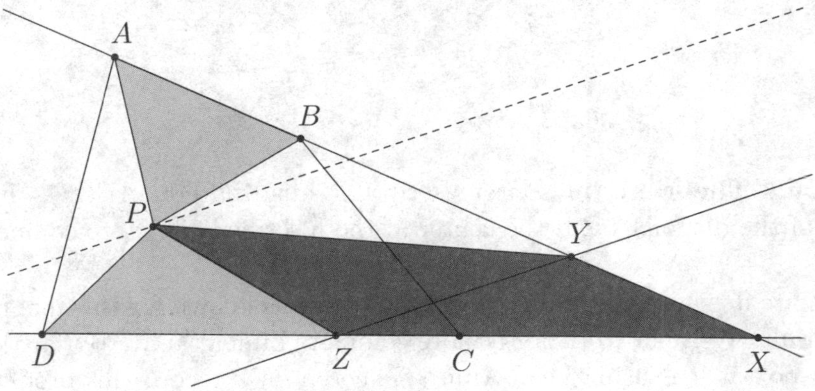

We note that

$$\begin{aligned} S_{PAB} + S_{PCD} &= S_{PXY} + S_{PXZ} \\ &= S_{PZXY} \\ &= S_{PYZ} + S_{XYZ}. \end{aligned}$$

Triangle XYZ is however fixed, as all points X, Y, Z are fixed (by construction). Thus, S_{XYZ} is a constant, and so

$$S_{PXY} = (S_{PAB} + S_{PCD}) - S_{PXZ}$$

is a constant. But this just means that $XY \cdot \delta(P, XY)$ is constant, which yields immediately that the locus of P is a line (at constant distance from the fixed line XY).

■

Comment. This result is most often associated with of the most classical theorems in quadrilateral geometry:

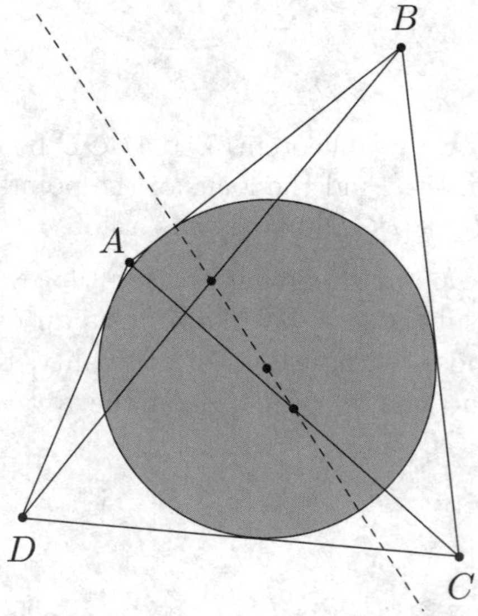

Newton's Theorem. In a circumscriptible quadrilateral $ABCD$, the midpoints of the diagonals and the center of the inscribed circle are collinear.

Despite its simplicity, there aren't many proofs known for this magnificent little result. We refer to [1] for a more complete bibliography. By far the best way to look at it is using Leon-Anne's Theorem. We record this proof below.

Proof. Note that

$$[MAB] + [MCD] = \frac{1}{2}[ABC] + \frac{1}{2}[ADC] = \frac{1}{2}[ABCD],$$

since M is the midpoint of AC. Similarly, we get that

$$[NAB] + [NCD] = \frac{1}{2}[ABD] + \frac{1}{2}[BCD] = \frac{1}{2}[ABCD].$$

Furthermore,

$$\begin{aligned}[IAB] + [ICD] &= \frac{1}{2}r \cdot (AB + CD) \\ &= \frac{1}{4}r \cdot (AB + CD + DA + BC) \\ &= \frac{1}{2}([IAB] + [IBC] + [ICD] + [IDA]) \\ &= \frac{1}{2}[ABCD],\end{aligned}$$

where the second equality holds by the well known (and easy to prove) Pithot's theorem that a quadrilateral $ABCD$ is circumscriptile if and only if

$$AB + CD = AD + BC.$$

Problem 107. (Titu Andreescu, Luis Gonzalez and Cosmin Pohoata, Mathematical Reflections) Let $ABCD$ be a convex quadrilateral and let M, N be the midpoints of diagonals AC and BD, respectively. Let $XYZT$ be the quadrilateral determined by the internal angle bisectors of its four angles. Prove that $XYZT$ is cyclic and that points M, N and the anticenter of $XYZT$ are collinear.

Solution. The key idea is to use Leon-Anne's Theorem, as we did when we proved Newton's theorem. Before that, however, we will need however another important preliminary result.

Lemma. Let $ABCD$ be a cyclic quadrilateral and let H_A, H_B, H_C, H_D be the orthocenters of triangles BCD, CDA, DAB and ABC, respectively. Then, the segments AH_A, BH_B, CH_C, DH_D all have the same midpoint.

Proof of Lemma. It clearly suffices to show that segments AH_A and BH_B have the same midpoint. To do this, we will argue that A, B, H_A and H_B are the vertices of a parallelogram.

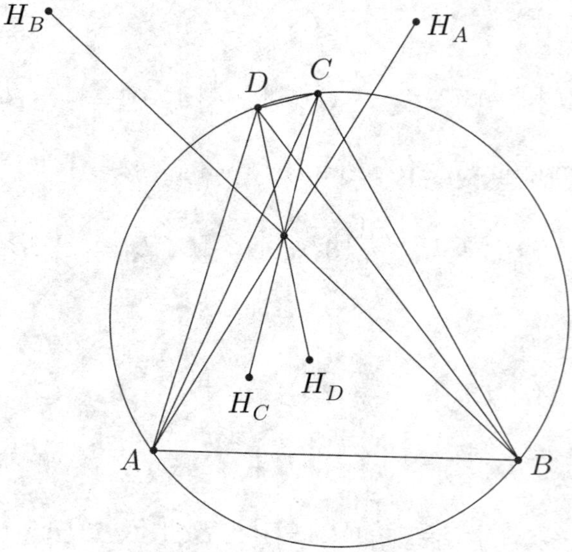

Indeed, note that AH_B and BH_A are parallel, since the lines are both perpendicular to CD. Also, as segments, they are both equal in length, since $AH_B = 2R|\cos CAD| = 2R|\cos CBD| = BH_A$, where R is the circumradius of the cyclic quadrilateral $ABCD$. This proves that ABH_AH_B is a parallelogram, as claimed.

The common midpoint of AH_A, BH_B, CH_C, DH_D described above is known in literature as the *anticenter* of (the cyclic) quadrilateral $ABCD$ and it has magnificient other properties. For example, we will see in Problem 95 that this point is also the common point of the nine-point circles of triangles BCD, CDA, DAB and ABC. There are many interesting alternative ways to describe this point, but for our purposes, the result from the above Lemma will be enough.

Returning to our problem, we see immediately that $XYZT$ is cyclic. Just note that

$$\begin{aligned}
\angle YXT &= 180^\circ - (\angle XDC + \angle XCD) \\
&= 180^\circ - (\frac{1}{2}\angle ADC + \frac{1}{2}\angle BCD) \\
&= 180^\circ - (180^\circ - \frac{1}{2}\angle BAD - \frac{1}{2}\angle ABC) \\
&= \frac{1}{2}\angle BAD + \frac{1}{2}\angle ABC \\
&= \angle ZAB + \angle ZBA \\
&= 180^\circ - \angle YZT.
\end{aligned}$$

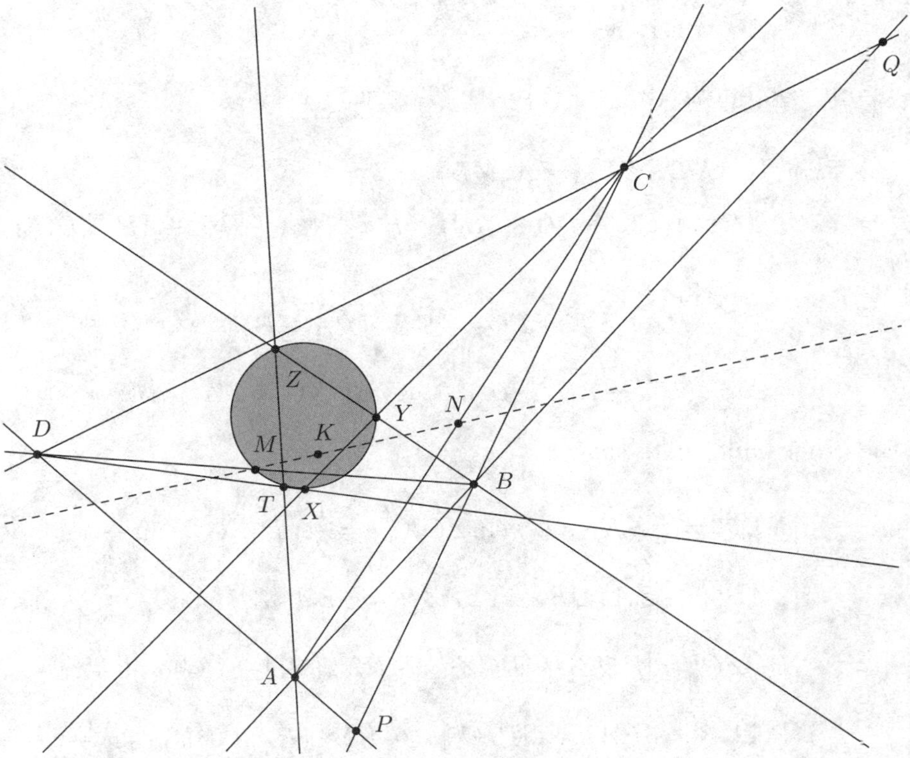

It is for the second part that we have to do all the work. Let P be the intersection of BC and AD and let Q be the intersection of AB and CD. Without loss of generality, we consider the configuration where X and T are the incenters of PCD and QDA and Y, Z are the excenters of QBC, PAB againts vertices Q and P, respectively (as in the diagram).

For convenience, label $\angle P = \angle APB$. In triangles PAB and PCD, recall the well known identities

$$\cot\frac{P}{2} = \frac{PZ \cdot AB}{ZA \cdot ZB}, \quad \cot\frac{P}{2} = \frac{PX \cdot CD}{XD \cdot XC}.$$

It follows that

$$\begin{aligned}
&\cot\frac{P}{2} \cdot (XD \cdot XC + ZA \cdot ZB) \\
&= PZ \cdot AB + PX \cdot DC \\
&= PZ \cdot \left(2\cos\frac{P}{2} \cdot PX + AB\right) - PX \cdot \left(2\cos\frac{P}{2} \cdot PZ - DC\right). \qquad (1)
\end{aligned}$$

Also, recall that in triangles PAB and PCD we have that

$$\cos\frac{P}{2} = \frac{PA + PB + AB}{2 \cdot PZ} \qquad (2)$$

$$\cos\frac{P}{2} = \frac{PD + PC - DC}{2 \cdot PX}. \tag{3}$$

Therefore, combining (2) and (3) with (1) gives us

$$\begin{aligned} &\cot\frac{P}{2} \cdot (XD \cdot XC + ZA \cdot ZB) \\ = \; &PZ \cdot (PD + PC - DC + AB) - PX \cdot (PA + PB + AB - DC); \end{aligned}$$

hence

$$\begin{aligned} &PZ \cdot (PD + PC) - PX \cdot (PA + PB) \\ &= XZ \cdot (DC - AB) + \cot\frac{P}{2} \cdot (XD \cdot XC + ZA \cdot ZB). \end{aligned} \tag{4}$$

On the other hand, note that

$$\begin{aligned} [XDC] &= \frac{1}{2}XD \cdot XC \cdot \sin\angle CXD = \frac{1}{2}XD \cdot XC \cdot \sin\frac{P}{2}, \\ [ZAB] &= \frac{1}{2}ZA \cdot ZB \cdot \sin\angle AZB = \frac{1}{2}ZA \cdot ZB \cdot \sin\frac{P}{2}, \\ [PDZC] &= [PDZ] + [PCZ] = \frac{1}{2}PZ \cdot (PD + PC) \cdot \sin\frac{P}{2}, \\ [PAXB] &= [PAX] + [PBX] = \frac{1}{2}PX \cdot (PA + PB) \cdot \sin\frac{P}{2}. \end{aligned}$$

Combining these four latter expressions with (4) yields

$$\begin{aligned} &\frac{1}{2}XZ \cdot (DC - AB) \cdot \sin\frac{P}{2} + [XDC] + [ZAB] \\ &= [PDZC] - [PAXB] \\ &= [ABCD] - [XAB] - [ZCD]. \end{aligned} \tag{5}$$

Let E, F be the orthocenters of triangles XYT and ZYT. Then, according to Theorem 2, $EXZF$ is a parallelogram with diagonal intersection $K = XF \cap ZE$, the anticenter of (cyclic) quadrilateral $XYZT$. The angle between EX and CD clearly equals $\angle XYZ$, and similarly the angle between ZF and AB equals $\angle XYZ$. Hence

$$[DECX] = [EDC] - [XDC] = \frac{1}{2}EX \cdot DC \cdot \sin XYZ$$

$$[AFBZ] = [ZAB] - [FAB] = \frac{1}{2}EX \cdot AB \cdot \sin XYZ;$$

thus

$$[EDC] + [FAB] = \frac{1}{2}EX(DC - AB) \cdot \sin XYZ + [XDC] + [ZAB].$$

But

$$EX = XZ \cdot \frac{\cos YZT}{\sin XYZ} = XZ \cdot \frac{\sin \frac{P}{2}}{\sin XYZ};$$

hence

$$[EDC] + [FAB] = \frac{1}{2} XZ(DC - AB) \cdot \sin \frac{P}{2} + [XDC] + [ZAB] \qquad (6)$$

From (5) and (6) we get

$$[EDC] + [FAB] + [XAB] + [ZCD] = [ABCD]. \qquad (7)$$

Now, since K is the midpoint of FX and EZ, we get

$$2[KAB] = [XAB] + [FAB] \ , \ 2[KDC] = [EDC] + [ZDC],$$

which yields

$$2[KAB] + 2[KDC] = [XAB] + [FAB] + [EDC] + [ZDC]. \qquad (8)$$

Finally, from (7) and (8), we conclude that

$$[KAB] + [KCD] = \frac{1}{2}[ABCD].$$

Therefore, by the Lemma, we conclude that K is on the line MN of the convex $ABCD$. This completes the proof.

■

Comment. Clearly, when $ABCD$ is circumscriptible, the internal angle bisectors of the angles of the quadrilateral are concurrent at the center of the inscribed circle; thus $X = Y = Z = T = K$, and so we recover Newton's classical theorem.

Problem 108. (Titu Andreescu and Cosmin Pohoata, USAMO 2013) Let ABC be a triangle. Find all points P on segment BC with the following property: If X, Y are the intersections of line PA with the common external tangents of the circumcircles of triangles PAB and PAC, then

$$\left(\frac{PA}{XY}\right)^2 + \frac{PB \cdot PC}{AB \cdot AC} = 1.$$

Solution. Let $\Gamma_1(O_1, R_1)$ and $\Gamma_2(O_2, R_2)$ be the circumcircles of triangles PAB and PAC. We begin with a lemma.

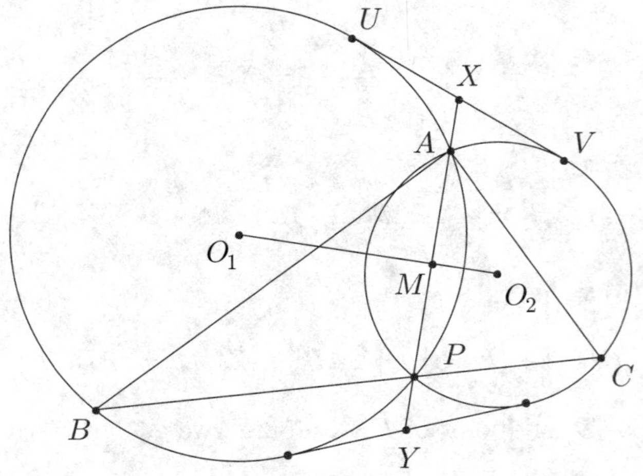

Lemma. Let U and V be the endpoints of one common external tangent of Γ_1 and Γ_2 (viewed as a segment). Then, we have that

$$XY = UV \cdot \frac{R_1 + R_2}{O_1O_2}.$$

Proof. Assume without loss of generality that UV is the common external tangent that contains X. It suffices to show that

$$XY \cdot O_1O_2 = UV \cdot (R_1 + R_2).$$

In this form, note that the left hand side is nothing but four times the area of triangle XO_1O_2, whereas the right hand side is twice the sum of the areas of triangles UVO_1 and UVO_2. However, if W is the intersection of O_1X with O_2V, triangles O_1XU and WXV are congruent, so

$$[UVO_1] + [UVO_2] = 2[UXO_1] + 2[VXO_2] = 2[WXO_2],$$

where the first equality holds from the fact that X is the midpoint of UV (power of point). Now, just notice that $[WXO_2] = [XO_1O_2]$, since X is also the midpoint of O_1XW by construction; thus, the desired relation is proven. □

Keeping the additional notations from the Lemma, notice that triangles O_1AO_2 and BAC are similar. Indeed, we always have that $\angle O_1AO_2 = \angle BAC$ and $\frac{AB}{AO_1} = \sin \angle APB = \sin \angle APC = \frac{AC}{AO_2}$ (the Law of Sines), hence the similarity.

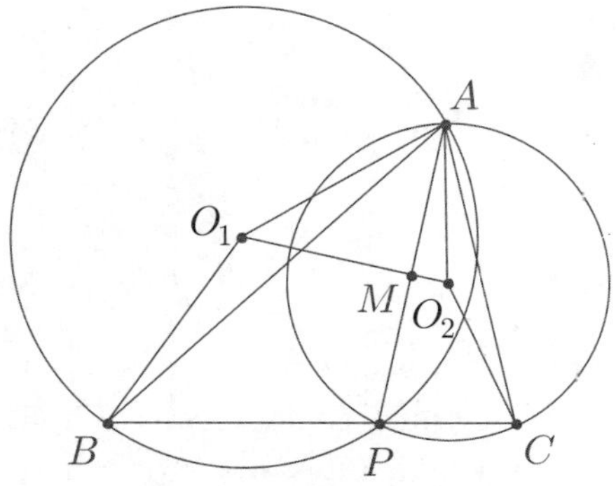

Thus, we can write

$$\frac{R_1}{AB} = \frac{R_2}{AC} = \frac{O_1O_2}{BC} = \frac{AP}{2h_a},$$

where the latter holds because AP is the A-altitude in triangle O_1AO_2. Moreover, if s denotes the semiperimeter of ABC, we note that

$$\frac{AP}{2h_a} = \frac{R \cdot AP}{bc} = \frac{UV}{2\sqrt{(s-b)(s-c)}},$$

where the latter now holds because of the similarity and the fact that

$$UV^2 = O_1O_2^2 - (R_1 - R_2)^2 = (O_1O_2 - R_1 + R_2)(O_1O_2 + R_1 - R_2).$$

Now, we turn to the Lemma. Combining it with what we found above for UV, we get

$$XY^2 = UV^2 \cdot \left(\frac{R_1 + R_2}{O_1O_2}\right)^2 = (s-b)(s-c) \cdot \frac{R^2 \cdot AP^2}{b^2c^2} \cdot \left(\frac{b+c}{a}\right)^2,$$

and so, using the well known equalities $abc = 4R \cdot [ABC]$ and

$$s(s-a)(s-b)(s-c) = [ABC]^2,$$

we deduce that

$$XY^2 = \frac{4(s-b)(s-c)R^2 \cdot AP^2 \cdot (b+c)^2}{16R^2[ABC]^2} = \frac{AP^2(b+c)^2}{4s(s-a)}.$$

Hence,

$$\left(\frac{AP}{XY}\right)^2 = \frac{4s(s-a)}{(b+c)^2}.$$

In other words, we have that points P lying on BC satisfying

$$\left(\frac{AP}{XY}\right)^2 + \frac{PB \cdot PC}{AB \cdot AC} = 1$$

need to satisfy

$$\frac{PB \cdot PC}{AB \cdot AC} = 1 - \frac{4s(s-a)}{(b+c)^2} = \frac{a^2}{(b+c)^2},$$

or equivalently,

$$PB \cdot PC = \frac{a^2bc}{(b+c)^2}.$$

There are precisely two points satisfying this condition: the foot of the A-internal angle bisector and its reflection across the midpoint of BC. Indeed, if P is the foot of the A-internal angle bisector, then it is well known that $PB = \frac{ac}{b+c}$ and $PC = \frac{ab}{b+c}$; hence, $PB \cdot PC = \frac{a^2bc}{(b+c)^2}$. Also, if Q is the reflection of P across the midpoint of BC, then $QB = PC$ and $QC = PB$, so $QB \cdot QC = PB \cdot PC = \frac{a^2bc}{(b+c)^2}$. In order to settle the question, it remains to show that we can't have more than two points satisfying the above equality. There are two ways to do this. First, notice that $PB \cdot PC$ is just the power of P with respect to the circumcircle of ABC, so $PB \cdot PC = |OP^2 - R^2|$, and so OP is constant; consequently, the only points P satisfying the above equality can only be the intersections of the line BC with the circle centered at O having radius the square root of the constant. Second, one could just write

$$\frac{a^2bc}{(b+c)^2} = PB \cdot PC = PB \cdot (a - PB),$$

and note that this gives a quadratic equation in PB, which can have at most two solutions. This completes the proof.

■

Problem 109. (Cosmin Pohoata, American Mathematical Monthly) Let ABC be a triangle with side lengths a, b, c, and area S and let x, y, z be three positive real numbers. Then,

$$a^2+b^2+c^2 \geq 4\sqrt{3}S + \frac{2}{x+y+z}\left(\frac{x^2-yz}{x}\cdot a^2 + \frac{y^2-zx}{y}\cdot b^2 + \frac{z^2-xy}{z}\cdot c^2\right).$$

Solution. We begin with the following preliminary result that lies at the heart of lots of difficult geometric inequalities. Its history is a little fuzzy, but geometers usually attribute it to John Conway.

Conway's Substitution Theorem. Let u, v, w be three reals such that the numbers $v+w$, $w+u$, $u+v$ and $vw+wu+uv$ are all nonnegative. Then, there exists a triangle XYZ with side lengths $x = YZ = \sqrt{v+w}$, $y = ZX = \sqrt{w+u}$, $z = XY = \sqrt{u+v}$. This triangle satisfies $y^2+z^2-x^2 = 2u$, $z^2+x^2-y^2 = 2v$, $x^2+y^2-z^2 = 2w$. The area $[XYZ]$ of this triangle equals $[XYZ] = \frac{1}{2}\sqrt{vw + wu + uv}$. If $X = \angle ZXY, Y = \angle XYZ, Z = \angle YZX$ are the angles of this triangle, then $\cot X = \frac{u}{2[XYZ]}$, $\cot Y = \frac{v}{2[XYZ]}$ and $\cot Z = \frac{w}{2[XYZ]}$.

Proof. Since the numbers $v + w$, $w + u$, $u + v$ are nonnegative, their square roots $\sqrt{v+w}$, $\sqrt{w+u}$, $\sqrt{u+v}$ exist, and, of course, are nonnegative as well. A straightforward computation shows that $\sqrt{w+u} + \sqrt{u+v} \geq \sqrt{v+w}$. Similarly, $\sqrt{u+v} + \sqrt{v+w} \geq \sqrt{w+u}$ and $\sqrt{v+w} + \sqrt{w+u} \geq \sqrt{u+v}$. Thus, there exists a triangle XYZ with side lengths

$$x = YZ = \sqrt{v+w},\ y = ZX = \sqrt{w+u},\ z = XY = \sqrt{u+v}.$$

It follows that

$$y^2 + z^2 - x^2 = \left(\sqrt{w+u}\right)^2 + \left(\sqrt{u+v}\right)^2 - \left(\sqrt{v+w}\right)^2 = 2u.$$

Similarly, $z^2 + x^2 - y^2 = 2v$ and $x^2 + y^2 - z^2 = 2w$. According now to the fact that

$$\cot Z = \frac{x^2 + y^2 - z^2}{4T},$$

we deduce that $\cot Z = \frac{w}{2[XYZ]}$, and similarly $\cot X = \frac{u}{2[XYZ]}$ and $\cot Y = \frac{v}{2[XYZ]}$. The well known trigonometric identity

$$\cot Y \cot Z + \cot Z \cot X + \cot X \cot Y = 1,$$

now becomes

$$\frac{v}{2[XYZ]} \cdot \frac{w}{2[XYZ]} + \frac{w}{2[XYZ]} \cdot \frac{u}{2[XYZ]} + \frac{u}{2[XYZ]} \cdot \frac{v}{2[XYZ]} = 1$$

or

$$vw + wu + uv = 4[XYZ]^2,$$

i.e.,

$$T = \frac{1}{2}\sqrt{4[XYZ]^2} = \frac{1}{2}\sqrt{vw + wu + uv}.$$

This completes the proof. □

Next, we prove the following partial result that could easily be a problem by itself in a contest.

Claim. Let ABC be a triangle with side lengths a, b, c, and area $[ABC]$, and let u, v, w be three reals such that the numbers $v + w$, $w + u$, $u + v$ and $vw + wu + uv$ are all nonnegative. Then,

$$ua^2 + vb^2 + wc^2 \geq 4\sqrt{vw + wu + uv} \cdot [ABC].$$

Using the Conway substitution theorem, we can construct a triangle with side lengths $x = \sqrt{v+w}$, $y = \sqrt{w+u}$, $z = \sqrt{u+v}$ and area $[XYZ] = \sqrt{vw+wu+uv}/2$. Let this triangle be XYZ. In this case, by the Neuberg-Pedoe inequality of Problem 30, applied for triangles ABC and XYZ, we get that

$$a^2\left(y^2+z^2-x^2\right)+b^2\left(z^2+x^2-y^2\right)+c^2\left(x^2+y^2-z^2\right) \geq 16[ABC][XYZ].$$

By the formulas from Conway's theorem, this becomes equivalent with

$$a^2 \cdot 2u + b^2 \cdot 2v + c^2 \cdot 2w \geq 16[ABC] \cdot \frac{1}{2}\sqrt{vw+wu+uv}$$

which simplifies to $ua^2 + vb^2 + wc^2 \geq 4\sqrt{vw+wu+uv} \cdot [ABC]$, which proves our Claim.

Returning to the problem, let $m = xyz(x+y+z) - 2yz(x^2 - yz)$, $n = xyz(x+y+z) - 2zx(y^2 - zx)$, and $p = xyz(x+y+z) - 2xy(z^2 - xy)$. The three terms $n+p$, $p+m$, and $m+n$ are all positive, and since

$$mn + np + pm = 3x^2y^2z^2(x+y+z)^2 \geq 0,$$

by the Claim above, we get that

$$\sum_{\text{cyclic}} [xyz(x+y+z) - 2yz(x^2-yz)]a^2 \geq 4\sqrt{3}xyz(x+y+z)[ABC].$$

This rewrites as

$$\sum_{\text{cyclic}} \left[(x+y+z) - 2\cdot\frac{x^2-yz}{x}\right] a^2 \geq 4\sqrt{3}(x+y+z)[ABC],$$

and, thus,

$$a^2+b^2+c^2 \geq 4\sqrt{3}[ABC]+\frac{2}{x+y+z}\left(\frac{x^2-yz}{x}\cdot a^2+\frac{y^2-zx}{y}\cdot b^2+\frac{z^2-xy}{z}\cdot c^2\right).$$

This completes the proof.

■

Comment. Plugging in $x = a$, $y = b$, $z = c$ in the original inequality and using the fact that

$$a^3+b^3+c^3-3abc = \frac{1}{2}(a+b+c)\left[(a-b)^2+(b-c)^2+(c-a)^2\right],$$

the problems implies the celebrated Hadwiger-Finsler inequality that

$$a^2+b^2+c^2 \geq 4\sqrt{3}[ABC] + (a-b)^2 + (b-c)^2 + (c-a)^2.$$

Problem 110. (Yiu's Theorem) Let ABC be a triangle and let D_1 ,D_2 be the tangency points of the $A-$excircle with the lines AB and AC, respectively. Similarly, define B_1, B_1, and C_1, C_2. Furthermore, let E_1E_2 and F_1F_1 meet at X, F_1F_2 and D_1D_2 meet at Y, and D_1D_2 and E_1E_2 meet at Z. Prove that the points X, Y, Z lie on the altitudes of triangle ABC.

Solution. This problem requires the reader to know an entire arsenal of properties about the excircles of the triangle in order to come up with the following solution. For sake of self-containment, we will go slowly through all of these, taking advantage of this magnificent theorem to review a lot of interesting properties of this important configuration.

We begin with the following observation.

Lemma 1. The foot of the perpendicular from B to the external angle bisector of angle ACB lies on the line D_1D_2.

This is very expected to happen given the result from the Claim we showed in the solution of Problem 32. It is always possible to extend a result about the incircle and incenter to each of the excircles. Consequently, a similar proof using harmonic divisions is possible. We shall however give a different argument.

Proof. Let X_1 be the foot of the perpendicular from B to the external angle bisector of angle C, X_2 the foot of the perpendicular from C to the external angle bisector of angle B, and I_a the A-excenter of triangle ABC. Since $\angle BX_1C = \angle CX_2B = 90°$, points X_1 and X_2 belong to the circle of diameter BC; in particular $\angle CBX_1 = \angle CX_2X_1$.

On the other hand, triangles CBX_1 and CI_aD_2 are similar; thus, $\angle CBX_1 = \angle CI_aD_2$. Moreover, quadrilateral $CX_2I_aD_1$ is cyclic (because of the opposite right angles at X_2 and D_2), so $\angle CI_aD_2 = \angle CX_2D_2$. It thus follows that $\angle CX_2X_1 = \angle CX_2D_2$, which means that points X_1, X_2, D_2 are collinear. Similarly, we get that points X_1, X_2, D_1 are collinear, and so we are done. This proves Lemma 1.

□

For the next Lemma, we define first in the same way the points Y_1, Y_2, Z_1, Z_2, and note that $Y_1, Y_2 \in E_1E_2$ and $Z_1, Z_2 \in F_1F_2$. The result that we need is the following beautiful

Lemma 2. Points X_1, X_2, Y_1, Y_2, Z_1, Z_2 are concyclic.

Proof. This is an immediate consequence of Taylor's theorem from Problem 4. Indeed, consider the excentral triangle $I_aI_bI_c$. In this triangle the lines

I_aA, I_bB, I_cC are, as we know, the altitudes, and by Lemma 1, the points X_1, X_2, Y_1, Y_2, Z_1, Z_2 are orthogonal projections of vertices A, B, C on the corresponding sides of $I_aI_bI_c$ (in some order). Taylor's theorem thus yields that they are concyclic. This proves Lemma 2.

□

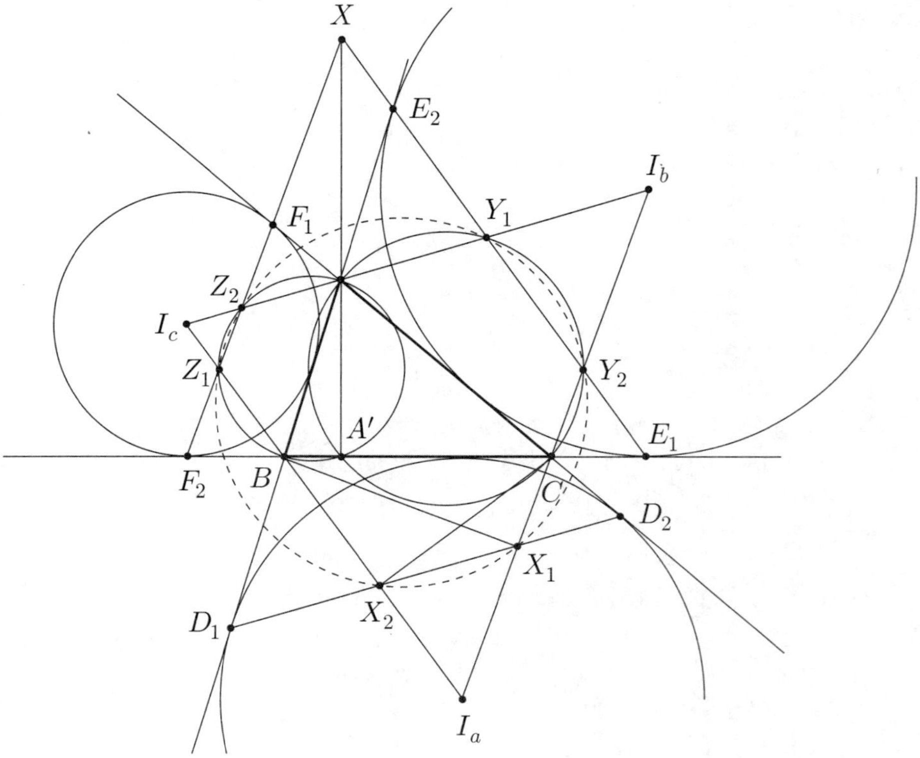

We are now ready to come back and prove the problem. Let Γ be the circle determined by points X_1, X_2, Y_1, Y_2, Z_1, Z_2, ω_c the circle of diameter AB, ω_b he circle of diameter AC, and A' the foot of the altitude from A in triangle ABC. By Lemmas 1 and 2, we have that Z_1Z_2 is the radical axis of Γ and ω_c, Y_1Y_2 is the radical axis of Γ and ω_b, and AA' is the radical axis of ω_b and ω_c. Thus, the three lines are concurrent at the radical center of Γ, ω_b, ω_c. In particular, this proves that X lies on the A-altitude AA' of triangle ABC. Similarly, we get that Y lies on BB' and that Z lies on CC', hence completing the proof.

■

References and Further Reading

1. Altschiller-Court, N., *College Geometry, an Introduction to the Modern Geometry of the Triangle and the Circle*, Dover publications, 2007.

2. Andreescu, T.; Feng, Z., *101 Problems in Algebra from the Training of the USA IMO Team*, Australian Mathematics Trust, 2001.

3. Andreescu, T.; Feng, Z., *102 Combinatorial Problems from the Training of the USA IMO Team*, Birkhäuser, 2002.

4. Andreescu, T.; Feng, Z., *103 Trigonometry Problems from the Training of the USA IMO Team*, Birkhäuser, 2004.

5. Andreescu, T.; Andrica, D.; Feng, Z., *104 Number Theory Problems from the Training of the USA IMO Team*, Birkhäuser, 2006.

6. Andreescu, T.; Feng, Z.; Loh, P., *USA and International Mathematical Olympiads 2004*, Mathematical Association of America, 2005.

7. Andreescu, T.; Feng, Z., *USA and International Mathematical Olympiads 2003*, Mathematical Association of America, 2004.

8. Andreescu, T.; Feng, Z., *USA and International Mathematical Olympiads 2002*, Mathematical Association of America, 2003.

9. Andreescu, T.; Feng, Z., *USA and International Mathematical Olympiads 2001*, Mathematical Association of America, 2002.

10. Andreescu, T.; Feng, Z., *USA and International Mathematical Olympiads 2000*, Mathematical Association of America, 2001.

11. Andreescu, T.; Feng, Z.; Lee, G.; Loh, P., *Mathematical Olympiads: Problems and Solutions from around the World, 2001–2002*, Mathematical Association of America, 2004.

12. Andreescu, T.; Feng, Z.; Lee, G., *Mathematical Olympiads: Problems and Solutions from around the World, 2000–2001*, Mathematical Association of America, 2003.

13. Andreescu, T.; Feng, Z., *Mathematical Olympiads: Problems and Solutions from around the World, 1999–2000*, Mathematical Association of America, 2002.

14. Andreescu, T.; Feng, Z., *Mathematical Olympiads: Problems and Solutions from around the World, 1998–1999*, Mathematical Association of America, 2000.

15. Andreescu, T.; Kedlaya, K., *Mathematical Contests 1997–1998: Olympiad Problems from around the World, with Solutions*, American Mathematics Competitions, 1999.

16. Andreescu, T.; Kedlaya, K., *Mathematical Contests 1996–1997: Olympiad Problems from around the World, with Solutions*, American Mathematics Competitions, 1998.

17. Andreescu, T.; Kedlaya, K.; Zeitz, P., *Mathematical Contests 1995–1996: Olympiad Problems from around the World, with Solutions*, American Mathematics Competitions, 1997.

18. Andreescu, T.; Enescu, B., *Mathematical Olympiad Treasures*, 2nd edition, Birkhäuser, 2011.

19. Andreescu, T.; Gelca, R., *Mathematical Olympiad Challenges*, 2nd edition, Birkhäuser, 2009.

20. Andreescu, T.; Andrica, D.; Cucurezeanu, I., *An Introduction to Diophantine Equations: A Problem-Based Approach*, Birkhäuser, 2010.

21. Andreescu, T.; Andrica, D., *360 Problems for Mathematical Contests*, GIL Publishing House, 2003.

22. Andreescu, T.; Andrica, D., *Complex Numbers from A to Z*, Birkhäuser, 2004.

23. Andreescu, T.; Feng, Z., *A Path to Combinatorics for Undergraduate Students: Counting Strategies*, Birkhäuser, 2003.

24. Andreescu, T.; Andrica, D., *Number Theory - A Problem Solving Approach*, Birkhäuser, 2009.

25. Ayme, J-L., *Sawayama and Thebault's Theorem*, Forum Geom. 3, 225-229, 2003.

26. Coxeter, H. S. M.; Greitzer, S. L., *Geometry Revisited*, New Mathematical Library, Vol. 19, Mathematical Association of America, 1967

27. Coxeter, H. S. M., *Non-Euclidean Geometry*, The Mathematical Association of America, 1998.

28. Doob, M., *The Canadian Mathematical Olympiad 1969–1993*, University of Toronto Press, 1993.

29. Engel, A., *Problem-Solving Strategies*, Problem Books in Mathematics, Springer, 1998.

30. Fomin, D.; Kirichenko, A., *Leningrad Mathematical Olympiads 1987–1991*, MathPro Press, 1994.

31. Fomin, D.; Genkin, S.; Itenberg, I., *Mathematical Circles*, American Mathematical Society, 1996.

32. Gelca, R.; Andreescu, T., *Putnam and Beyond*, Springer, 2007.

33. Graham, R. L.; Knuth, D. E.; Patashnik, O., *Concrete Mathematics*, Addison-Wesley, 1989.

34. Greitzer, S. L., *International Mathematical Olympiads, 1959–1977*, New Mathematical Library, Vol. 27, Mathematical Association of America, 1978.

35. Holton, D., *Let's Solve Some Math Problems*, A Canadian Mathematics Competition Publication, 1993.

36. Kazarinoff, N. D., *Geometric Inequalities*, New Mathematical Library, Vol. 4, Random House, 1961.

37. Kedlaya, K; Poonen, B.; Vakil, R., *The William Lowell Putnam Mathematical Competition 1985–2000*, The Mathematical Association of America, 2002.

38. Klamkin, M., *International Mathematical Olympiads, 1978–1985*, New Mathematical Library, Vol. 31, Mathematical Association of America, 1986.

39. Klamkin, M., *USA Mathematical Olympiads, 1972–1986*, New Mathematical Library, Vol. 33, Mathematical Association of America, 1988.

40. Kürschák, J., *Hungarian Problem Book, volumes I & II*, New Mathematical Library, Vols. 11 & 12, Mathematical Association of America, 1967.

41. Kuczma, M., *144 Problems of the Austrian–Polish Mathematics Competition 1978–1993*, The Academic Distribution Center, 1994.

42. Kuczma, M., *International Mathematical Olympiads 1986–1999*, Mathematical Association of America, 2003.

43. Larson, L. C., *Problem-Solving Through Problems*, Springer-Verlag, 1983.

44. Lausch, H. *The Asian Pacific Mathematics Olympiad 1989–1993*, Australian Mathematics Trust, 1994.

45. Liu, A., *Chinese Mathematics Competitions and Olympiads 1981–1993*, Australian Mathematics Trust, 1998.

46. Liu, A., *Hungarian Problem Book III*, New Mathematical Library, Vol. 42, Mathematical Association of America, 2001.

47. Lozansky, E.; Rousseau, C. *Winning Solutions*, Springer, 1996.

48. Mitrinovic, D. S.; Pecaric, J. E.; Volonec, V. *Recent Advances in Geometric Inequalities*, Kluwer Academic Publisher, 1989.

49. Mordell, L.J., *Diophantine Equations*, Academic Press, London and New York, 1969.

50. Niven, I., Zuckerman, H.S., Montgomery, H.L., *An Introduction to the Theory of Numbers*, Fifth Edition, John Wiley & Sons, Inc., New York, Chichester, Brisbane, Toronto, Singapore, 1991.

51. Prasolov, V.V., *Problems in Plane Geometry*, Fifth Edition, Moscow textbooks, 2006.

52. Savchev, S.; Andreescu, T. *Mathematical Miniatures*, Anneli Lax New Mathematical Library, Vol. 43, Mathematical Association of America, 2002.

53. Sharygin, I. F., *Problems in Plane Geometry*, Mir, Moscow, 1988.

54. Sharygin, I. F., *Problems in Solid Geometry*, Mir, Moscow, 1986.

55. Shklarsky, D. O; Chentzov, N. N; Yaglom, I. M., *The USSR Olympiad Problem Book*, Freeman, 1962.

56. Slinko, A., *USSR Mathematical Olympiads 1989–1992*, Australian Mathematics Trust, 1997.

57. Szekely, G. J., *Contests in Higher Mathematics*, Springer-Verlag, 1996.

58. Tattersall, J.J., *Elementary Number Theory in Nine Chapters*, Cambridge University Press, 1999.

59. Taylor, P. J., *Tournament of Towns 1980–1984*, Australian Mathematics Trust, 1993.

60. Taylor, P. J., *Tournament of Towns 1984–1989*, Australian Mathematics Trust, 1992.

61. Taylor, P. J., *Tournament of Towns 1989–1993*, Australian Mathematics Trust, 1994.

62. Taylor, P. J.; Storozhev, A., *Tournament of Towns 1993–1997*, Australian Mathematics Trust, 1998.

63. Yaglom, I. M., *Geometric Transformations*, New Mathematical Library, Vol. 8, Random House, 1962.

64. Yaglom, I. M., *Geometric Transformations II*, New Mathematical Library, Vol. 21, Random House, 1968.

65. Yaglom, I. M., *Geometric Transformations III*, New Mathematical Library, Vol. 24, Random House, 1973.

Other Books from XYZ Press

1. Andreescu, T.; Ganesh, A., *109 Inequalities from the AwesomeMath Summer Program*, 2014.
2. Andreescu, T. *Mathematical Reflections - two great years*, 2014.
3. Becheanu, M., Enescu, B., *Balkan Mathematical Olympiads*, 2014.
4. Andreescu, T.; Branislav K., *Math Leads for Mathletes: A rich resource for young math enthusiasts, parents, teachers, and mentors*, 2014.
5. Andreescu, T.; Ganesh, A., *108 Algebra Problems from the AwesomeMath Year-Round Program*, 2014.
6. Andreescu, T., *Mathematical Reflections - two more years*, 2014.
7. Andreescu, T.; Rolinek, M.; Tkadlec, J., *107 Geometry Problems from the AwesomeMath Summer Program*, 2013.
8. Andreescu, T.; Rolinek, M.; Tkadlec, J., *106 Geometry Problems from the AwesomeMath Year-Round Program*, 2013.
9. Andreescu, T., *105 Geometry Problems from the AwesomeMath Year-Round Program*, 2013.
10. Andreescu, T.; Kane, J., *Purple Comet Math Meet! - the first ten years*, 2013.
11. Andreescu, T.; Dospinescu, G., *Straight from the Book*, 2012.
12. Andreescu, T.; Boreico, I.; Mushkarov, O.; Nikolov, N., *Topics in Functional Equations*, 2012.
13. Andreescu, T., *Mathematical Reflections - the next two years*, 2012.
14. Andreescu, T., *Mathematical Reflections - the first two years*, 2011.
15. Andreescu, T.; Dospinescu, G., *Problems from the Book*, 2008.

Printed by “Combinatul Poligrafic”
Com. nr. 41150